Solitary Waves in Plasmas and in the Atmosphere

Solitary Waves in Plasmas and in the Atmosphere

Vladimir Petviashvili
Kurchatov Atomic Energy Institute, Moscow, USSR

Oleg Pokhotelov
Earth Physics Institute, Moscow, USSR

Translated from the Russian

Routledge
Taylor & Francis Group
LONDON AND NEW YORK

Published 1992 by Gordon and Breach Science Publishers

Published 2016 by Routledge
2 Park Square, Milton Park, Abingdon, Oxon OX14 4RN
52 Vanderbilt Avenue, New York, NY 10017, USA

First issued in paperback 2019

Routledge is an imprint of the Taylor & Francis Group, an informa business

Library of Congress Cataloging-in-Publication Data

Petviashvili, V.I. (Vladimir Iosifovich)
[Uedinennye volny v plazme i atmosfere. English]
Solitary waves in plasmas and in the atmosphere / by V.I. Petviashvili and O.A. Pokhotelov.
p. cm.
Translation of: Uedinennye volny v plazme i atmosfere.
Includes bibliographical references and index.
ISBN 2-88124-787-3
1. Space plasmas 2. Solitons. 3. Magnetohydrodynamic waves.
I. Pokhotelov, O.A. (Oleg Aleksandrovich II. Title.
QC718.5.M36P4713 1992
530.4'4--dc20 91-27253
CIP

ISBN 13: 978-0-367-45043-4 (pbk)
ISBN 13: 978-2-88124-787-3 (hbk)

Contents

Preface

This book arose as a result of the authors' work on the review "Solitary Vortices in Plasmas" written for the *Soviet Journal of Plasma Physics*. While working on this review, we found that our subject is part of a rapidly developing discipline, namely, the theory of solitary waves. Since we have had some experience in the field, we thought it useful to write a unified exposition of both its principles and more recent advances. We believe that the most interesting properties of solitary waves can be adequately described in terms of ordinary mathematics at the customary level of rigor in the physical literature. Progress in the mathematical theory of solitary waves (the inverse scattering method, the Darboux transform, etc.) is largely a consequence of the complete integrability property for a comparatively narrow class of nonlinear equations, while most nonlinear equations and their solitary solutions of physical interest are outside that class. Numerous elegant and important results have nevertheless been obtained for them. The authors' interests center around plasma physics. However, since many results in this field may also be useful elsewhere, mainly in the physics of the ocean and atmosphere, it seemed natural to extend our treatment to include a wider range of topics. It turns out that some results in the physics of the ocean and atmosphere may be useful for the theory of nonlinear waves in plasmas. This theory has a short history. Sagdeev was the first in 1958 to discover solitary plasma waves in magnetosonic and ion acoustic modes (see e.g. Sagdeev, 1966). Zabusky and Kruskal (1965) later introduced the concept of solitons. Since then, the study of nonlinear waves and particularly a special case of them, namely solitary waves, has become a rapidly developing frontier of physics and mathematics. Some nonlinear equations of the evolutionary type have been found to be reducible to sets of linear equations, i.e., they appear to be implicitly linear. Nearly all of them have important applications in physics. They include the two-dimensional equation obtained by Kadomtsev and Petviashvili (1970). The properties of such equations and their integration using the method of the inverse scattering problem (ISP) have been described in detail by Zakharov *et al.* (1980). Valuable introductions to the theory of nonlinear plasma waves have been written by Karpman (1975), Kadomtsev (1976), Hasegawa (1975), Artsimovic and Sagdeev (1979) and Chen (1984). A similar book by Rabinovic and Trubetskov (1984) is devoted mainly to oscillations that are randomized as a result of nonlinearity. The present monograph is primarily concerned with vortex-like structures and their stability. We also include important results from laboratory modeling of plasma and atmospheric vortices in shallow rotating water by Antonova *et al.* Some of the evidence for their existence in the magnetosphere and in planetary atmospheres is reviewed.

The questions discussed here are mostly ones with which the authors have been directly involved. While not claiming comprehensive completeness, we have tried to deal with the most important aspects of the problem. Some papers of interest, published both in the USSR and abroad, may inadvertently have gone unnoticed. Some readers may feel a need for a more rigorous and detailed substantiation of some of our conclusions. The gaps can be filled in by the reader himself, thus serving as a useful substitute for problem-solving exercises.

The authors are grateful to Academician Sagdeev for helpful discussions and to colleagues with whom some of the results described here were obtained. They also express their sincere gratitude to all those who have been involved in the preparation of the manuscript.

Vladimir Petviashvili
Oleg Pokhotelov

Introduction

With the development of nonlinear wave theory some novel concepts came into use, such as solitary waves, solitary vortices, and solitons. There is still some inconsistency in the use of these terms. They all refer to solitary disturbances that do not change their shape with time. Some of these disturbances retain their shape because dispersive spreading is compensated by a nonlinear effect, namely, the frequency correlation of the Fourier components in a wave packet. Such disturbances may be stable if they minimize the Lyapunov functional – the sum of the energy integral and other integrals of the motion. They can be termed solitons in a dispersive medium. This name is frequently applied to isolated structures that conserve their shape owing to the invariance of certain topological properties: current streamlines being closed or coupled with one another (helicity), sets of magnetic surfaces embedded in one another, etc. Such structures are called topological solitons. Solitons of mixed types are also possible, for example, solitary vortices.

Dispersion is directly related to the shielding or screening of external disturbances by a medium (e.g., the Debye shielding of charges in a plasma). For this reason soliton size is characterized by the shielding length and velocity of propagation. When the velocity of a solitary disturbance is the same as that of a linear wave, this gives rise to radiation instead of screening. Generally the size of a soliton in a dispersive medium is smaller when the difference between its velocity and the velocity of the linear wave is larger. In some media linear waves with very small velocities exist (e.g., ion-acoustic and Alfvén waves across a magnetic field in a plasma, and Rossby waves in the atmosphere). Traveling vortices of small amplitude can then arise from them in which the shielding length is close to the characteristic dispersion size (which is equal to the ion cyclotron radius in a plasma or to the Rossby length in a rotating atmosphere).

It is now generally recognized that for media in which stable solitons can exist, an arbitrary initial disturbance that is localized in space will emit the excess energy in the form of free waves and will be transformed into a set of solitons during the characteristic screening time. Because the interaction among solitons is weak, the characteristic correlation length between disturbances in a medium of this sort is of the order of the size of a soliton. For this reason, statistical methods of study are appropriate here.

The study of solitary waves (solitons) should ultimately lead to a better understanding of the internal structure of elementary particles. That is why solitons are attracting great attention from physicists and mathematicians. The material presented in this book shows that solitons are also of interest in plasma physics and geophysics, where these concepts have many important applications. They are also fascinating as a purely mathematical object that is amenable to deep analysis.

Solutions of Euler's equation in hydrodynamics in the form of 2D and 3D solitary vortices were well-known as far back as the last century. These are lumps of kinetic energy that are not subject to spreading owing to local conservation of vorticity flux and other quantities. In recent years large-scale atmospheric vortices have been widely studied on the basis of the model equation put forward by Charney (1946) to describe Rossby waves. A remarkable property of that equation is that it has no one-dimensional analog, since it involves nonlinearity in the form of a two-dimensional vector product (Jacobian). Hasegawa and Mima (1978) showed that this equation describes potential drift waves in plasma, thus revealing a close analogy between drift waves and Rossby waves. This occurs because the Coriolis force in a rotating fluid and the Lorentz force in a magnetized plasma are similar. Experiments with vortices in shallow rotating water in a paraboloidal vessel provide new data for further study of this phenomenon (Antonova *et al.*, 1987).

The first paper on plasma vortices was by Kadomtsev and Pogutse (1974) who derived a simplified equation describing MHD vortices elongated along the magnetic field. Because of the smallness of the vortices compared with the typical dimensions of the plasma and the magnetic field, the distorting effect of the inhomogeneity is small and can be compensated by vortex pumping processes owing to reconnection processes (Kadomtsev, 1987).

The subsequent development of plasma vortex theory involved taking wave dispersion into account, which, along with the finiteness of the drift velocity, gives rise to qualitatively new effects. In particular, solitary vortices of the Alfvén type, from which dispersion effects play a significant role, can be spontaneously amplified as a result of electron dissipation. The free energy is replenished by the plasma inhomogeneity in this case (Petviashvili and Pokhotelov, 1986). It is important that, although the presence of shear in the magnetic field stabilized dissipative instabilities, solitary Alfvén vortices are not sensitive to shear because they are localized over small dimensions. They have soliton properties. In dispersive media the Fourier components that comprise a wave packet have different frequencies that are nonlinear functions of the wave vector in the linear approximation. As a result, in the course of time a wave is distorted and spreads out in coordinate space. In other words, it undergoes dispersion and diffraction. When nonlinearity is present, the Fourier components begin to interact, producing new components. This leads to correlations among components with different wave vectors and to such behavior as the breakdown or collapse of the wave packet. Thus, dispersion and diffraction counteract the nonlinear effects. Wave packets can develop in many modes of oscillation, in which these effects counter each other in a stable way. Naturally, a packet of this kind cannot have an arbitrary shape. It must be a solitary wave or a vortex of definite form that travels without change.

Solitary vortices are qualitatively different from solitary waves in that although they can travel relative to the medium, they carry particles that are rotating in the vortex (trapped particles), thus increasing the plasma thermal conductivity, diffusion, and electrical resistivity. Vortices develop in those modes that have a small phase velocity compared with the speed of sound, for example, drift Alfvén waves in plasmas and Rossby waves in the atmosphere.

Kosevich, Ivanov and Kovalev (1983) have described another important class of topological solitary vortices in spin waves in solids with a constant magnetization vector.

Structures should be investigated for stability. For this we mostly use the Lyapunov method. According to it, stability studies in elastic systems where all wave frequencies are real requires that the first integrals of the motion be available. Although this method is not easy to use, it has the merit of providing a sufficient criterion for stability. We note that spectral methods yield sufficient criterion for instability, so that by themselves they cannot solve the problem of whether a given solution can actually exist.

Our discussion is based on the derivation of a simplified (model) equation in which the main effects show up explicitly, while less important quantitative corrections are disregarded. The conservation of symmetry exhibited by the original equations is a necessary condition. In particular, analogs of integrals of motion that the original equations possess must be conserved. This obvious requirement is emphasized because there are papers that neglect to pay the attention due to it. Some authors use the simplified equations to obtain solutions that do not satisfy the necessary conditions of smoothness. Those solutions are sometimes of interest, but should be treated with caution.

At the same time, the concept of solitons has come to be used for solitary steady-state disturbances propagating in active dissipative media. One example is an impulse propagating along nerve fibers (Scott, 1970). These disturbances obey nonlinear equations, in which dispersion occurs through diffusion and other dissipative effects. In contrast to this case, dissipative effects can be disregarded to a first approximation in the equations of plasmas and atmospheres. They can be included as corrections through perturbation theory. The Langmuir caviton is an exception. In recent years a theory of thermodynamic equilibrium between solitons and free waves has been developed by Yankov that is analogous to the equilibrium theory for vapor-liquid systems.

1. Some results from the linear theory

1.1. LEONTOVICH'S PARABOLIC APPROXIMATION

The equations for plasmas and the atmosphere are rather unwieldy in the general case and even in the linear approximation. This is because they have a large number of collective degrees of freedom corresponding to different oscillation modes. There are important limiting cases where small parameters are available, in which these equations can be simplified by expanding in powers of the small parameters, since it is possible to separate one mode from the others. A popular method of simplification is the so called parabolic approximation put forward by Leontovich (1944). This can be applied to a hyperbolic 3D linear equation of the type

$$\partial_t^2\varphi-\nabla^2\varphi+\omega_0^2\varphi=-\alpha\nabla^4\varphi, \tag{1.1}$$

where ω_0 and α are constants. The case $\omega_0=0$ corresponds to acoustic waves, and $\omega_0\neq 0$ to optical waves. The constant α represents a dispersion parameter and can be either positive or negative. The sign of α defines a medium as one of positive or negative dispersion.

In the long-wavelength limit, the equation can be reduced to one of the parabolic type.

We first consider acoustic oscillations ($\omega_0=0$). We shall seek φ as a function of time and position of the form

$$\varphi=\varphi(z-t,x,y,t). \tag{1.2}$$

Assume the dependence of φ on the first argument to be much stronger than on the rest. This amounts to saying that the wave packet propagates mainly along the z-axis. We also assume that the dispersion effects are small, i.e., the right-hand side of (1.1) is small compared with the others. Substituting (1.2) into (1.1) and discarding small terms, we get

$$\partial_z(2\partial_t \varphi - \alpha\partial_z^3\varphi) + \nabla_\perp^2\varphi = 0; \ \partial_z \gg \nabla_\perp \tag{1.3}$$

The derivative ∂_t is taken with respect to the last argument in (1.2). While retaining the main properties of the original (1.1), this new equation describes waves that travel in the direction of increasing and decreasing z, independently of one another.

In the case of optical waves simplification is possible if the third term in (1.1) is large compared to the second and the right-hand side can be neglected. We then seek φ as a function of position and time in the form

$$\varphi = \frac{1}{2}\psi(r,t,)\exp\left(-i(\omega_0^2 + k_0^2)^{1/2}t + ik_0z\right) + c.c.\,, \tag{1.4}$$

where $c.c.$ denotes the complex conjugate. The complex amplitude ψ is assumed to be weakly dependent on time. In that case, (1.4) describes a wave packet having frequency ω_0, wavenumber k_0, and complex envelope ψ. Substituting (1.4) into the original equation (1.1) and discarding small terms, we get

$$2i(\omega_0^2 + k_0^2)^{1/2}\partial_t\psi + 2ik_0\partial_z\psi + \nabla_\perp^2\psi = 0\,. \tag{1.5}$$

This kind of simplification is like the transition from the Klein–Gordon equation to Schrödinger's nonrelativistic equation in quantum theory. When the energy density of the oscillations is not very small, equations (1.3) and (1.5) can be supplemented with a simple nonlinear term. The result will be either the Kadomtsev–Petviashvili nonlinear equation or Schrödinger's nonlinear equation.

1.2. WAVES IN UNIFORM PLASMAS

1.2.1. Ion acoustic waves

As is well known, there is a great variety of oscillations and waves in plasma. We may conveniently begin our account with the simplest oscillations of the acoustic type. Ion acoustic waves in a homogeneous plasma are an example of this type of oscillation.

We shall consider the propagation of these waves when the plasma pressure is small compared to the pressure of the magnetic field, $\beta \equiv 8\pi \frac{p_0}{B_0^2} << 1$ (p_0 is the unperturbed plasma pressure, B_0 is the unperturbed magnetic field). The electric field E can be regarded as having a potential; i.e., curl E $= -\partial_t \mathbf{B} \simeq 0$. This means that it can be represented in the form $\mathbf{E} = -\nabla\varphi$, where φ is the electric potential. Ion acoustic oscillations exist only in a nonisothermal plasma where $T_e >> T_i$, i.e., when the electron temperature is large compared with the ion temperature. If that condition is not satisfied, the wave velocity $c_s = (T_e / m_i)^{1/2}$ approaches the ion thermal speed $v_{T_i} = (2T_i / m_e)^{1/2}$. Ion Landau damping is strong in that case. Because the ion acoustic velocity is small compared to the electron thermal speed, the electron number density has the time to approach a Boltzmann distribution

$$n_e = n_0 \exp(e\varphi / T_e) \tag{1.6}$$

owing to the mobility of the electrons.

The equations of motion and continuity for the ions in the collisionless case have the form (in the linear approximation)

$$\partial_t \mathbf{v} = -(e/m_i)\nabla\varphi + \omega_{Bi}(e_z \times \mathbf{v}), \tag{1.7}$$

$$\partial_t n_i + n_0 \operatorname{div} \mathbf{v} = 0, \quad \omega_{Bi} = eB_0/m_i c. \tag{1.8}$$

Here $\mathbf{v}$ is the hydrodynamic velocity, n_0 is the unperturbed particle number density, n_i is the perturbed part of ion density, and $e_z = \mathbf{B}_0/B_0$ is the unit vector along the magnetic field. The ion pressure is assumed to be small; hence, its gradient can be neglected in the equation of motion.

With small wavelengths it is important that charge neutrality is violated, as described by Poisson's equation

$$\nabla^2\varphi = 4\pi e(n_e - n_i). \tag{1.9}$$

The left-hand side of this equation is a measure of the extent to which neutrality is violated and leads to dispersion with a typical scale length of $r_0 = (T_e / 4\pi n_0 e^2)^{1/2}$, the Debye length.

A wave packet that is localized in time and space can be represented as a Fourier integral

$$\varphi(\mathbf{r},t) = \int \psi(\mathbf{k},\omega) \exp(i\mathbf{kr} - i\,\omega t)\, d\mathbf{k}\, d\omega\ , \tag{1.10}$$

where $\psi(\mathbf{k},\omega)$ is a function that is localized in wavenumber and frequency space. In the linear approximation, substituting (1.10) into (1.7) to (1.9), we find that ψ is not a smooth function of ω, but has the form $\psi = \psi_{\mathbf{k}}\delta(\omega - \omega_{\mathbf{k}})$, where $\omega_{\mathbf{k}}$ is a function of $\mathbf{k}$ and is given by the dispersion relation

$$1 + k^2 r_D^2 = \frac{k_z^2 c_s^2}{\omega_{\mathbf{k}}^2} - \frac{k_\perp^2 r_s^2\, \omega_{Bi}^2}{\omega_{Bi}^2 - \omega_{\mathbf{k}}^2}\ , \tag{1.11}$$

δ (x) is Dirac's delta function, and $r_s = c_s / \omega_{Bi}$ is the ion Larmor radius determined by the electron temperature.

Here the packet spreads more slowly than would be the case if ψ were a smooth function of ω. We know that a packet of the form (1.10) does not change shape only if $\omega_{\mathbf{k}}$ is a linear function of $\mathbf{k}$,

$$\omega_k = \omega_0 + \mathbf{b}\cdot\mathbf{k}, \tag{1.12}$$

where ω_0 and $\mathbf{b}$ are constants. A packet of this kind is referred to as a steady-state packet. $\omega_0 = 0$ in an acoustic wave. Below we shall be interested in wave packets for which the departures from (1.12) are small. The departures from (1.12) can be classified as dispersive or diffractive. Their significance is determined by the size of the localization region of the carrier $\psi_{\mathbf{k}}$. If the localization is strong enough, one can restrict oneself to the first few terms of the expansion. The corrections along $\mathbf{b}$ are referred to as dispersive and those across it as diffractive. This distinction is quite clear in the isotropic case, where the effect of the magnetic field can be neglected $\omega \gg \omega_{Bi}$. Consider an almost one-dimensional packet in which the localization along $\mathbf{b}$ is much stronger than in the other directions. Take the z-axis along $\mathbf{b}$. Then the expression for $\omega_{\mathbf{k}}$ can be expanded into

$$\omega_k \simeq c_s k(1 - k^2 r_D^2) - c_s k_z (1 - k_z^2 r_D^2/2 + k_\perp^2/2k_z^2). \tag{1.13}$$

This relation corresponds to the parabolic approximation in coordinate space and leads to the equation

$$\partial_z(\partial_t \Phi + c_s \partial_z \Phi + (c_s r_D^2/2)\partial_z^3 \Phi) = -(c_s/2)\nabla_\perp^2 \Phi;$$

$$\Phi \equiv e\varphi / T_e \tag{1.14}$$

which shows clearly that the packet travels along z, and undergoes dispersive spreading along z and diffraction in the x and y directions.

It is customary to speak of media with positive and negative dispersion. In the 1D case, dispersion is referred to as positive if the phase velocity ω/k increases with wavenumber, and negative if it decreases.

It follows from (1.13) and (1.14) that it is also meaningful to speak of media with positive and negative diffraction for more than one dimension. Note that negative diffraction can occur in a given direction only in anisotropic media. Anisotropy appears in ion acoustic waves at low frequencies such that $\omega \lesssim \omega_{Bi}$. In the limit $\omega \ll \omega_{Bi}$ we have from (1.11) that

$$\omega \simeq c_s k_z(1 - k_z^2 r_D^2/2 - k_\perp^2 r_s^2/2). \tag{1.15}$$

It follows from (1.15) that diffraction of ion acoustic waves is negative for low frequencies and $k_z \gg k_\perp$.

A linear equation for a packet in the coordinate space that is similar to (1.14) is obtained from (1.15) and has the form

$$\partial_t \Phi + c_s \partial_z \Phi + (c_s r_D^2/2)\partial_z^3 \Phi = -(c_s r_s^2/2)\nabla_\perp^2 \partial_z \Phi. \tag{1.16}$$

It follows from (1.15) and (1.16) that the z component of the phase velocity for linear ion acoustic waves is always below c_s. We shall see below that this means that, although the corresponding solitary 1D wave travels faster than c_s, it is stable in an isotropic plasma (without a magnetic field) and unstable in an anisotropic plasma (with a magnetic field). To sum up, stability is controlled by the ratio of the signs of the dispersion and diffraction, rather than by the diffraction sign alone.

1.2.2. Alfvén and magnetosonic waves

Ion acoustic oscillations propagate only in a strongly nonisothermal plasma with $T_e \gg T_i$. In a magnetized plasma, oscillations of the same frequencies can propagate in which the electric field does not have a potential and a significant role is played by oscillations in the magnetic field. These oscillations are weakly attenuated, even in the isothermal case, because their velocity is large compared with v_{T_i}. The relevant dispersion relation can be obtained by using the equations of two-fluid hydrodynamics. We neglect the thermal corrections for now, incorporating them where necessary. In the linear approximation we have

$$m_i\partial_t \mathbf{v}_i = e\mathbf{E} + (e/c)(\mathbf{v}_i\times\mathbf{B}_0), \tag{1.17}$$

$$m_e\partial_t \mathbf{v}_e = -e\mathbf{E} - (e/c)(\mathbf{v}_e\times\mathbf{B}_0), \tag{1.18}$$

$$\partial_t n_{i,e} + n_0 \operatorname{div} \mathbf{v}_{i,e} = 0, \tag{1.19}$$

$$\operatorname{curl} \mathbf{B} = \frac{4\pi e n_0}{c}(\mathbf{v}_i - \mathbf{v}_e), \tag{1.20}$$

$$\partial_t \mathbf{B} = -c\cdot\operatorname{curl} \mathbf{E}. \tag{1.21}$$

We shall be interested in frequencies below the ion plasma frequency ω_{pi}. In that case, the oscillations can be assumed to be quasineutral, i.e.,

$$n_i \simeq n_e = n. \tag{1.22}$$

Defining the mass velocity of the plasma as

$$\mathbf{u} = (m_i\mathbf{v}_i + m_e\mathbf{v}_e)/m, \quad m \equiv m_i + m_e, \tag{1.23}$$

adding (1.17) and (1.18), and using (1.19) and (1.21), we obtain

$$\partial_t \mathbf{u} = \frac{1}{4\pi\rho_0}(\operatorname{curl} \mathbf{B}\times\mathbf{B}_0). \tag{1.24}$$

When the inertia of the ions and electrons is taken into account, the electric field has the form

$$\mathbf{E} = -\frac{1}{c}(\mathbf{u}\times\mathbf{B}_0) + \frac{m_i}{e}\partial_t\mathbf{u} - \frac{m_e}{e}\partial_t\mathbf{v}_e. \tag{1.25}$$

Maxwell's equations (1.20), (1.21) then become

$$\operatorname{curl} \mathbf{B} = \frac{4\pi e n_0}{c}\frac{m}{m_i}(\mathbf{u} - \mathbf{v}_e), \tag{1.26}$$

$$\partial_t \mathbf{B} = \operatorname{curl}(\mathbf{u}\times\mathbf{B}_0) - \frac{B_0}{\omega_{Bi}}\partial_t\mathbf{u} + \frac{B_0 m_e}{m_i\omega_{Bi}}\partial_t\mathbf{v}_e. \tag{1.27}$$

Equations (1.24), (1.26), (1.27) together with the equation of continuity

$$\partial_t n + n_0 \operatorname{div} \mathbf{u} = 0 \tag{1.28}$$

yield the following dispersion law

$$\omega_{1,2} = \frac{c_A k}{2s}\left\{\left[(1+\cos\theta)^2 + \frac{k^2 r_A^2}{s^2}\cos^2\theta\right]^{1/2} \pm \left[(1-\cos\theta)^2 + \frac{k^2 r_A^2}{s^2}\cos^2\theta\right]^{1/2}\right\}, \tag{1.29}$$

where $r_A = c_A/\omega_{Bi}$, $s^2 = 1 + k^2 r_0^2$, $r_0 = c/\omega_{pe}$, and θ is the angle between the wave vector and the magnetic field. The upper sign corresponds to magnetic sound, the lower to the Alfvén wave. For the case of small wave numbers $kr_A \ll 1$ (as $r_0 \ll r_A$, we also have $kr_0 \ll 1$). The dispersion relations for these modes take the well-known forms

$$\omega_1 = c_A k, \quad \omega_2 = c_A k \cos\theta. \tag{1.30}$$

When propagation takes place at a small angle with respect to the magnetic field $(k_\perp/k_z)^2 \ll k_z r_A$ and at wavelengths large compared to r_A, we find from (1.29) that

$$\omega = c_A k_z\left(1 \pm \left|k_z\right| r_A/2 + k_\perp^2/4k_z^2\right). \tag{1.31}$$

The electric field vector of the magnetosonic waves rotates in the same sense as the electron cyclotron rotation, and that of the Alfvén waves as that of the ion cyclotron rotation.

When the angle of propagation is not very small $(\theta^2 \ll k_z r_A)$ and $k_z r_A \ll 1$, equation (1.29) can be simplified to

$$\omega = c_A k\left[1 - \frac{k^2 r_A^2}{2}\left(\frac{m_e}{m_i} - \cot^2\theta\right)\right], \tag{1.32}$$

for magnetosonic waves and to

$$\omega = c_A k_z\left[1 - \frac{k^2 r_A^2}{2}\left(\frac{m_e}{m_i} + \cot^2\theta\right)\right], \tag{1.33}$$

for Alfvén waves.

One can see from (1.32), (1.33) that when the angle is close to $\pi/2$, the magnetosonic wave dispersion changes sign, while the Alfvén dispersion (if the thermal corrections are disregarded) remains the same. The characteristic dispersion scale length becomes of the order of skin length c/ω_{pe}. Therefore, when $\beta > m_e/m_i$ ($r_s > r_0$), the dispersion of these waves may be controlled by the thermal corrections that were omitted in our derivation of the dispersion relation (1.29). Equations (1.32), (1.33) then become

$$\omega = c_A k \left[1 - \frac{k^2 r_A^2}{2} \left(\frac{m_e}{m_i} - \cot^2 \theta + \beta_i/2 \right) \right] \tag{1.34}$$

and

$$\omega = c_A k_z \left[1 - \frac{k^2 r_A^2}{2} \left(\frac{m_e}{m_i} + \cot^2 \theta - \frac{\beta_e + (3/4)\beta_i}{2} \right) \right], \tag{1.35}$$

where $\beta_{e,i} = 8\pi n_0 T_{e,i}/B_0^2$ and $k_z \ll k_\perp$.

One can see from (1.34) that thermal corrections extend the region of angles where the magnetic sound has negative dispersion, whereas they change the sign of the dispersion of the Alfvén wave at $\beta > m_e/m_i$. The β_i term in the dispersion relations for the magnetosonic and Alfvén waves corresponds to the effect of a finite ion Larmor radius. Apart from this effect, the electron thermal corrections proportional to β_e also contribute to the dispersion of the Alfvén waves. This corresponds to including a finite longitudinal electric field $E_\parallel/E_\perp \simeq k^2 r_s^2$ in the Alfvén waves. Waves in which thermal corrections to dispersion are significant are called kinetic Alfvén or magnetosonic waves (Hasegawa, 1976). Note that when a magnetosonic wave propagates at a large angle with respect to the magnetic field, the dispersion remains weak, even for $\omega \gtrsim \omega_{Bi}$.

1.3. DISPERSION RELATIONS MODIFIED TO INCLUDE THE EFFECT OF PLASMA INHOMOGENEITIES

We now consider the effect of a small plasma inhomogeneity on the dispersion properties of the waves. An inhomogeneity is regarded as small if the Larmor radius of the particles is small compared with the characteristic scale length of the inhomogeneity a. In a state of equilibrium, a gradient in the plasma pressure gives rise to a diamagnetic current and associated electron and ion drift velocities $v_{ni,e} = \pm\, cT_{i,e}\varkappa_n/eB_0$

change in the dispersion properties of the waves, if the phase velocity is close to the drift velocity. Otherwise, the effect of inhomogeneities can be accomodated within the framework of the ordinary quasi-classical approximation. From this and the above remarks we conclude that inhomogeneities have a strong effect only on ion acoustic and Alfvén oscillations, whose phase velocity decreases when the angle between the wave vector and the magnetic field increases.

We first discuss the effect of inhomogeneities on ion acoustic waves. Incorporating the plasma inhomogeneity into their dispersion relation, for $T_i \to 0$ we obtain

$$1 + k_\perp^2 r_s^2 - \frac{\omega_{ne}}{\omega} - \frac{k_z^2 c_s^2}{\omega^2} + \frac{i\sqrt{\pi}}{|k_z| v_{Te}} (\omega - \omega_{ne}) = 0, \qquad (1.36)$$

where $\omega_{ne} = k_y v_{ne}$ is the electron drift frequency. As k_z decreases , the phase velocity of these waves along the magnetic field increases and approaches the Alfvén velocity. The electric field of the oscillations continues to have a potential while $\omega/k_z \ll c_A$. If $\omega/k_z \gg c_s$ and no Landau damping is assumed, equation (1.36) simplifies to

$$\omega = \omega_{ne}(1 - k_\perp^2 r_s^2 + k_z^2 c_s^2/\omega_{ne}^2) + \frac{i\sqrt{\pi}\,\omega_{ne}^2}{|k_z| v_{Te}} (k_\perp^2 r_s^2 - k_z^2 c_s^2/\omega_{ne}^2) \qquad (1.37)$$

$$c_s \ll \omega/k_z \ll v_{Te}, c_A; \quad k_\perp r_s \ll 1.$$

Such waves are called potential drift waves. In this limit their propagation velocity along the magnetic field, ω/k_z, is greater than the ion thermal velocity, even in an isothermal plasma. For this reason drift potential waves are virtually free of the Landau ion damping, in contrast to ion sound. Another peculiar feature is that the velocity of propagation ω/k becomes less than the drift velocity. As one can see from (1.36), this changes the sign of the Landau damping. The result is the so-called dissipative drift instability (Timofeev, 1964; Moiseev and Sagdeev, 1964). The amplitude of the potential drift waves grows, although energy is dissipated. In some cases, collisional dissipation exceeds the Landau damping. The sign of the dissipation term can be shown to coincide with that of $\omega - \omega_{ne}$. Note that the soliton vortices of this mode discussed in Chapter 6 have a velocity higher than the drift velocity so that they cannot be amplified as a result of this effect. The longitudinal propagation velocity approaches the Alfvén velocity as k_z decreases. The perturbation that of $\omega - \omega_{ne}$. Note that the soliton vortices of this mode discussed in Chapter 6 have a velocity higher than the drift velocity so that they cannot be amplified as a result of this effect. The longitudinal propagation velocity approaches the Alfvén velocity as k_z decreases. The perturbation of the magnetic field then becomes important. The potential drift wave transforms into a drift Alfvén wave (Figure 1.1) with a dispersion relation of the form

$$(\omega^2 + \omega\omega_{ni} - k_z^2 c_A^2)(\omega - \omega_{ne})\left(1 + i\sqrt{\pi}\,\frac{\omega}{|k_z| v_{Te}}\right)$$
$$= k_\perp^2\, r_s^2 k_z^2\, c_A^2(\omega - \omega_{ni}). \qquad (1.38)$$

Here $\omega_{ni,e} = k_y v_{ni,e}$ are drift frequencies for the ions and electrons, respectively. If $\omega \simeq k_z c_A \lesssim \omega_{ne}$, then the Landau damping changes sign, leading to instability (Mikhailovskii and Rudakov, 1963). Instability can also occur due to electron collisional dissipation.

Drift waves are also possible when $k_\perp r_{B_i} >> 1$, where r_{B_i} is the ion Larmor radius. In this case the ions are not effectively magnetized and they obey the Boltzmann distribution. The dispersion relation then has the form (Mikhailovskii, 1974)

$$(\omega - \omega_{ni})(\omega - \omega_{ne}) = k_\perp^2 r_{Bi}^2\, k_z^2 c_A^2(1 + T_e/T_i). \qquad (1.39)$$

Equation (1.39) describes two modes that are travelling in the directions of the electron or ion Larmor drifts. The velocities of both modes exceed the respective drift velocities, so that they are stable in the linear approximation. As we shall see in Chapter 6, however, vortices of these modes travel at a lower velocity than the drift velocity. This makes them grow due to dissipation, as in the case of the drift dissipative instability.

When k_z is extremely small and $k_\perp r_{B_i} \ll 1$, the drift wave velocity may be greater than the Alfvén velocity and the electron thermal velocity. Electron inertia can be disregarded as before. These waves are known as flute waves. Because of the large wavelength along $\mathbf{B}_0$, the effects of curvature and inhomogeneity in the magnetic field are quite significant for these waves. In the simplest case, these effects can be taken into account by introducing an effective "force of gravity" that acts on the electrons and ions. Oscillations of the magnetic field can be neglected. Their dispersion relation has the form (Longmire and Rosenbluth, 1957)

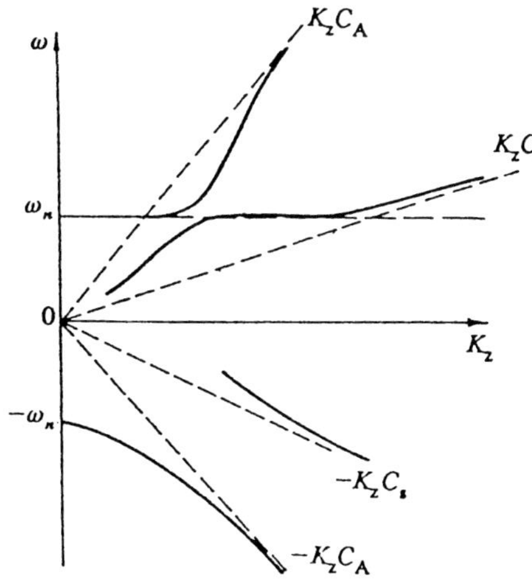

Figure 1.1. The frequencies of Alfvén and ion-acoustic oscillations in an inhomogeneous plasma as functions of the longitudinal wavenumber k_z with $k_\perp$ constant. It can be seen that propagation in the direction of the electron Larmor drift makes the frequencies of the two branches converge.

$$\omega^2 - \omega\omega_{ni} - g\varkappa = 0, \tag{1.40}$$

where g is the effective "gravity" given by $g \simeq P_0/\rho_0 R$, where P_0 is the plasma pressure, R is the characteristic scale length of the magnetic field inhomogeneity, $\varkappa = \partial_x \ln n_0$ is the characteristic scale length of the plasma inhomogeneity, and ρ_0 the mass density of the plasma.

1.4. FLUTE INSTABILITY IN THE EARTH'S MAGNETOSPHERE

The curvature of the lines of force was taken into account by defining an effective force of gravity in the preceding section. That approach is an oversimplification, since it neglects the effects of plasma compressibility that are observed in a real magnetic field. Here we shall incorporate these effects in more detail when we analyse the flute instability. In the absence of shear, this instability is the most dangerous one in magnetic confinement systems from the standpoint of rearranging the entire equilibrium configuration. It arises when the plasma pressure falls off rapidly enough with distance from the centre. The criterion for this kind of instability in confinement systems with closed lines of force was established by B.B. Kadomtsev and has the form (Kadomtsev, 1966)

$$V\partial_V p < -\gamma p, \tag{1.41}$$

where p is plasma pressure and $V = \int dl/B$ is the specific volume of a magnetic tube of force, with the contour integral being taken along a line of force. It is assumed in this derivation that the pressure is constant on the V = const surfaces; that is the pressure is a function of V alone. In a number of cases, for example, in the Earth's magnetosphere, however, the surfaces of equal p and V may be different. An equilibrium electric current then develops along **B**. The magnetosphere can enter such a state when plasma convection occurs from the magnetospheric tail (Akasofu and Chapman, 1972). We shall show that when the plasma pressure is not constant over isosurfaces, instabilities that were not known may occur (Ivanov and Pokhotelov, 1987).

Consider a magnetic confinement system with conducting ends. In the Earth's magnetosphere the role of conducting ends is played by a conducting ionosphere that has Pedersen's integral conductivity Σ_p (Akasofu and Chapman, 1972). The longitudinal current density flowing into the ionosphere is related to the divergence of the transverse electric field in the ionosphere by means of

$$j_{\|} = \Sigma_p \operatorname{div} \mathbf{E}_{\perp}, \quad \Sigma_p = \int \sigma_p dl, \tag{1.42}$$

where the integration is taken over the entire thickness of the ionosphere and σ_p is Pedersen's conductivity. The contribution of the magnetosphere to the longitudinal current densities is determined by the equation for the shorting of the currents,

$$\operatorname{div} \mathbf{j}_{\perp} + \operatorname{div} \mathbf{j}_{\|} = 0, \tag{1.43}$$

in which the transverse current is given in terms of the ion and electron densities and velocities in the magnetosphere by $\mathbf{j}_{\perp} = en(\mathbf{v}_{\perp i} - \mathbf{v}_{\perp e})$. The velocity of the particle motion and the particle density can be found by using the equations of two-fluid hydrodynamics (the subscript indicating the kind of particle is omitted)

$$\partial_t n + \operatorname{div} n\mathbf{v} = 0, \quad d_t \equiv \partial_t + \mathbf{v}\cdot\nabla, \tag{1.44}$$

$$mnd_t\mathbf{v} + \nabla\cdot\pi = -\nabla p + en(\mathbf{E} + c^{-1}(\mathbf{v}\times\mathbf{B})) \tag{1.45}$$

$$d_t p + \gamma p \operatorname{div} \mathbf{v} + (\gamma - 1) \operatorname{div} \mathbf{q} = 0. \tag{1.46}$$

Here π, q are the viscosity tensor and the heat flux. The heat flux across the magnetic field is taken in the form (Braginsky, 1965)

$$\mathbf{q}_\perp = \frac{5}{2}\frac{p}{m\omega_B}(\mathbf{e}_z\times\nabla T), \quad \mathbf{e}_z = \mathbf{B}/B. \tag{1.47}$$

Assuming that the frequency of the waves is lowto the ion cyclotron frequency, from the equation of motion (1.43) we obtain an expression for the ion and electron velocities in the drift approximation

$$\mathbf{v} = \mathbf{v}_E + \mathbf{v}_L + \mathbf{v}_I + \mathbf{v}_{\|}, \tag{1.48}$$

where $\mathbf{v}_E = (c/B^2)(\mathbf{E}\times\mathbf{B})$ is the electric drift, $\mathbf{v}_L = (c/eBn)(\mathbf{e}_z\times\nabla p)$ is the Larmor drift, and $\mathbf{v}_I = \omega_B^{-1}(\mathbf{e}_z\times d_t(\mathbf{v}_E + \mathbf{v}_L))$ is the polarization drift. Using these expressions for the ion and electron velocities, we find the transverse current to be

$$\mathbf{j}_\perp = \frac{c}{B}(\mathbf{e}_z\times\nabla(p_e + p_i)) + \frac{en}{\omega_{Bi}}(\mathbf{e}_z\times d_t(\mathbf{v}_E + \mathbf{v}_L)) \tag{1.49}$$

Substituting (1.49) into the current closure conditions (1.43) and modifying the equation of continuity for the electrons and the heat balance equation for ions and electrons by including (1.47), (1.48), we arrive at the following set of equations:

$$\operatorname{div}\left\{\frac{c}{B}\left(\mathbf{e}_z\times\nabla(p_e + p_i)\right)\right\} + \operatorname{div}\left\{\frac{en}{\omega_{Bi}}\left(\mathbf{e}_z\times d_t(\mathbf{v}_E + \mathbf{v}_L)\right)\right\} + \operatorname{div}\mathbf{j}_{\|} = 0; \tag{1.50}$$

$$d_t n + n\operatorname{div}\mathbf{v}_E + \operatorname{div} n\mathbf{v}_{ne} - \operatorname{div}\left(\frac{c}{enB}(\mathbf{e}_z\times\nabla p_e)\right) = 0; \tag{1.51}$$

$$d_t p_i + \gamma p_i\operatorname{div}\mathbf{v}_E + \gamma p_i\operatorname{div}\mathbf{v}_{\|i} + \gamma\operatorname{div}\mathbf{q}_{\|i} + \gamma\operatorname{div}\left(\frac{c}{eB}(\mathbf{e}_z\times\nabla(p_i T_i))\right) = 0; \tag{1.52}$$

$$d_t p_e + \gamma p_e\operatorname{div}\mathbf{v}_E + \gamma p_e\operatorname{div}\mathbf{v}_{\|e} + \gamma\operatorname{div}\mathbf{q}_{\|e} - \gamma\operatorname{div}\left(\frac{c}{eB}(\mathbf{e}_z\times\nabla(p_e T_e))\right) = 0. \tag{1.53}$$

We shall be interested in flute perturbations ($\mathbf{B}\cdot\nabla = 0$), which in the case of a low-β plasma ($\beta \ll 1$) involve the motion of magnetic flux tubes across the magnetic field without distortions. Flute oscillations in a curvilinear field are conveniently investigated by transforming to the Euler coordinates x^1, x^2, x^3; the magnetic field then has the form

$$\mathbf{B} = \nabla x^1 \times \nabla x^2. \tag{1.54}$$

These coordinates leave a single nonzero contravariant component, $B^3 = g^{-1/2}$, where g is the determinant of the metric tensor g_{ik} while the value of B is given by $B = (g_{33}B^3B^3)^{1/2} = (g_{33}/g)^{1/2}$. The coordinates x^1, x^2 describe plasma motion in directions across the magnetic field, provided $g_{13} = g_{23} = 0$, because in that case the x^1- and x^2-axes are orthogonal to the x^3-axis. For this reason all the variables in flute perturbations, apart from the magnetic field itself, are functions of the transverse coordinates x^1, x^2 only. Using the definition of the specific volume V, one can obtain an expression for it in terms of the Euler coordinates as an integral along a line of force, $V = \oint \sqrt{g}\,dx^3$. Perturbations of the electric field can be assumed to have a potential ($\mathbf{E} = -\nabla\varphi$) in a low-pressure plasma.

Transforming to the variables x^1, x^2, x^3 in the system of equations (1.50)–(1.53) and integrating along a magnetic field line, we obtain the averaged system of equations;

$$\{V, p_e + p_i\} + m_i n c d_t \Delta_2 \varphi + \frac{m_i c}{e} d_t \Delta_2 p_i = \frac{1}{c}(\sqrt{g} j_{i\parallel}) \Bigg|_{x^3 = c_1}^{x^3 = c_2}, \tag{1.55}$$

$$V\partial_t n + c\{\varphi, Vn\} + \frac{c}{e}\{V, p_i\} = -n(\sqrt{g}v_{\parallel e}) \Bigg|_{x^3 = c_1}^{x^3 = c^2}, \tag{1.56}$$

$$V\partial_t p_i + cV^{1-\gamma}\{\varphi, p_i V^\gamma\} - \gamma\frac{c}{e}\{V, p_i T_i\} = -\gamma p_i(\sqrt{g}v_{\parallel i}) \Bigg|_{x^3 = c_1}^{x^3 = c_2}, \tag{1.57}$$

and

$$V\partial_t p_e + cV^{1-\gamma}\{\varphi, p_e V^\gamma\} + \gamma\frac{c}{e}\{V, p_e T_e\}$$

$$= -\gamma p_e(\sqrt{g}v_{||e})\Bigg|_{x^3 = c_1}^{x^3 = c_2}; \quad d_t \equiv \partial_t + c\{\varphi, \ldots\} \tag{1.58}$$

The following notation has been used; $V_{ik} \equiv \int g_{ik}\sqrt{g}dx^3$, $\{f, g\} \equiv \partial_1 f\partial_2 g - \partial_2 f\partial_1 g$, $\Delta_2 \equiv V_{22}\partial_1^2 + V_{11}\partial_2^2$, $\partial_1 \equiv \partial_{x^1}$, and $\partial_2 \equiv \partial_{x^2}$. $j_{||}$, $v_{||e}$, and $v_{||i}$ are the contravariant components of the longitudinal current and of the longitudinal electron and ion velocities. The longitudinal current and the other quantities on the right-hand sides of (1.55)–(1.58) are taken at the ends of a line of force at the boundary with the ionosphere. The longitudinal heat flux at the boundary between the ionosphere and magnetosphere is assumed to be small.

Equation (1.55) yields a relation between the equilibrium pressure and the longitudinal equilibrium current:

$$c\{V, p\} = (\sqrt{g}j_{||})|_{x^3=c_2} - (\sqrt{g}j_{||})|_{x^3=c_1}, \tag{1.59}$$

where $p = p_{oe} + p_{oi}$ is the total unperturbed plasma pressure. When the boundary conditions are symmetric, (1.59) is transformed into the basic equation for the magnetosphere–ionosphere interaction derived by Tverskoi (Tverskoi, 1982).

A longitudinal current is generated because the total transverse current, with density $\mathbf{j}_\perp = (c/B)(\mathbf{e}_z \times \nabla p)$, which flows into a flux tube is not equal to the total transverse current that flows out of it when the equilibrium pressure is not constant at V = const surfaces. The resulting excess charge flows along the flux tube into the ionosphere.

We now proceed to discuss the system of equations (1.55)–(1.58) which describes flute perturbations. For the sake of completeness we shall derive the growth rate of the ordinary flute instability, when the pressure is constant on the V=const surfaces. Assuming the time and position dependence of the perturbed quantities has the form $\exp(-i\omega t + i\int k_1 dx^1 + i\int k_2 dx^2)$ and neglecting drift effects, we obtain

$$\omega^2 = \frac{1}{m_i n_0 k_\perp^2 V^\gamma}(\mathbf{k}\times\nabla V)(\mathbf{k}\times\nabla(pV^\gamma)),$$

$$\mathbf{A}\times\nabla B \equiv A_1\partial_2 B - A_2\partial_1 B, \tag{1.60}$$

where $k_\perp^2 = V_{22}k_1^2 + V_{11}k_2^2$. We assume for concreteness that the equilibrium quantities p and V are functions of x^1 only. Then (1.60) yields

$$\omega^2 = \frac{k_2^2\partial_1 V}{m_i n_0 k_\perp^2}(\partial_1 p + \gamma p V^{-1}\partial_1 V) = -\Gamma^2, \tag{1.61}$$

where Γ is the growth rate for the flute instability. (1.61) gives the criterion (1.41), according to which the pressure falls off fairly rapidly. In the case of a dipole field, this corresponds to having the plasma pressure, fall off with distance from the trap centre as $r^{-20/3}$ (Kadomtsev, 1962).

Now consider the case in which the criterion (1.41) is not satisfied. We shall show that plasma may be unstable even in that case, if the surfaces of constant pressure do not coincide with the V=const surfaces. By analogy with the instability discussed above, p and V are assumed to depend on x^1, but the pressure gradient is also assumed to have a component along the x^2-axis, which is related to the longitudinal current by $\partial_2 p\partial_1 V = 2\sqrt{g}j_{||}/c$ according to (1.59). Equations (1.55)–(1.58) yield the dispersion relation for flute oscillations in the presence of a longitudinal equilibrium current when drift effects are neglected:

$$\omega = -i\frac{\Sigma_p(k_\perp')^2}{2m_i n_0 k_\perp^2 c^2} \pm i\left[\frac{\Sigma_p(k_\perp')^2}{2m_i n_0 k_\perp^2 c^2} + \frac{2k_1k_2\sqrt{g}}{m_i n_0 k_\perp^2 c}j_{||} + \Gamma^2\right]^{1/2} \tag{1.62}$$

where $(k_\perp')^2 = V^{11}k_1^2 + V^{22}k_2^2$ and $V^{ik} \equiv g^{ik}\sqrt{g}$. Two limiting cases will be considered for (1.62). When the conductivity Σ_p is large, the relevant root of (1.62) can be written in the form

$$\omega = i\frac{m_i n_0 k_\perp^2 c^2}{\Sigma_p(k_\perp')^2}\left(\frac{2k_1k_2\sqrt{g}}{m_i n_0 k_\perp^2 c}j_{||} + \Gamma^2\right), \tag{1.63}$$

from which it follows that a magnetic flux tube may be unstable under interchange motions for $\Gamma^2 < 0$. In the other limiting case, when the conductivity is sufficiently small the inertia of particles in a magnetic flux tube should be included, so that (1.62) takes the form

$$\omega^2 = -\left(\frac{2k_1k_2\sqrt{g}}{m_in_0k_\perp^2c}j_\parallel + \Gamma^2\right). \qquad (1.64)$$

The maximum growth rate of the instability in the case of large longitudinal currents,

$$\gamma = \left(\pm\frac{\sqrt{g}}{cm_in_0(V_{11}V_{22})^{1/2}}j_\parallel + \Gamma^2\right)^{1/2}, \qquad (1.65)$$

is attained when $k_2 = \pm(V_{22}/V_{11})^{1/2}k_1$. The plus or minus sign in (1.65) is chosen depending on the direction of the longitudinal current. If the longitudinal current flows toward greater values of $x^3(j_{||} > 0)$, then the projections of the wave vector on the x^1- and x^2-axes satisfy the condition $k_1k_2 > 0$; otherwise, when the current is negative, the relation $k_1k_2 < 0$ holds.

If the criterion for the ordinary flute instability is not satisfied, that is, $\Gamma^2 < 0$, then a flute instability with variable pressure on the surfaces of equal specific volume cannot develop unless the longitudinal current exceeds some threshold value. The threshold can be estimated by using the expression for the growth rate (1.65):

$$j_{||}^* = cm_in_0(V_{11}V_{22})^{1/2}|\Gamma^2|/\sqrt{g}. \qquad (1.66)$$

When the wavelengths of the plasma perturbations are short enough, extra terms should be inserted in the dispersion relations derived above. To do this, we consider the complete set of equations (1.55)–(1.58) assuming low conductivity at the ends. For simplicity we first put $T_e \gg T_i$ and neglect the ion pressure perturbation $\tilde{p}_i = 0$. Linearization yields a dispersion relation for flute oscillations that incorporates magnetic drift effects:

$$\omega^2 - 2\gamma\omega_{Ve}\omega + \gamma\omega_{Ve}^2 + (1 - \omega_{Ve}/\omega)(\Gamma_j^2 + \Gamma^2)$$

$$- \gamma\omega_{Ve}\Omega_e^2/\omega = 0, \qquad (1.67)$$

where $\omega_{Ve} = cT_ek_2\partial_1V/eV$ is the magnetic drift electron frequency,

$$\Gamma_j^2 = \frac{2k_1k_2\sqrt{g}}{m_in_0k_\perp^2c}j_\parallel \quad \text{and}$$

$$\Omega_e^2 = \frac{k_2^2}{m_i k_\perp^2}\,[(\gamma - 1)T_e(\partial_1 V)^2/V + \partial_1 V \partial_1 T_e]. \tag{1.68}$$

(1.68) is derived using the condition $\partial_2 T_e = 0$, which follows from the heat balance equation for the electrons with the equilibrium quantities. An analysis of (1.67) shows that the flute instability does not develop for the magnetic drift frequencies $|\omega_{Ve}|^2 \gtrsim \Gamma_j^2 + \Gamma^2$. Comparing the value of ω_{Ve} and Γ, we see that stabilization of the flute instability occurs for wavelengths shorter than r_s

$$(k_\perp B/V)^{1/2} r_s \gtrsim \max(1, (R/a)^{1/2}), \tag{1.69}$$

where a is the typical scale length of plasma inhomogeneity and R is the radius of curvature of a magnetic field line. Stabilization of the instability for $\Gamma_j^2 \gg \Gamma^2$ takes place at still shorter wavelengths than those given by (1.69). The threshold value of the longitudinal current for short wavelength perturbations is greater than that for long wavelength perturbations given by (1.66), as one can see from the condition for the flute instability $\Gamma_j^2 + \Gamma^2 \gtrsim |\omega_{Ve}|^2$. We shall obtain the solution of the dispersion relation (1.67) for $\Gamma_j^2 + \Gamma^2 >> |\omega_{Ve}|^2$. In that case, the root of interest can be written as

$$\omega = \frac{1}{2}\gamma\omega_{Ve}\left(1 - \frac{\Omega_e^2}{\Gamma_j^2 + \Gamma^2}\right) + (\Gamma_j^2 + \Gamma^2)^{1/2}. \tag{1.70}$$

Equation (1.70) is a refinement of the solution of the dispersion relation (1.64) for the case of short wavelength flute disturbances. It follows from (1.70) that perturbations of this sort are oscillatory in character with a frequency on the order of the electron magnetic drift frequency.

Simple analytical expressions are also obtained for the case $T_i \gg T_e$. Linearizing the set of equations (1.55)–(1.58) and putting $\tilde{p}_e = 0$, we find the dispersion relation for flute oscillations:

$$(\omega - \gamma\omega_{Vi})(\omega - \omega_{Pi}) - \gamma\omega_{Vi}\,\omega_{Ti} - \gamma\omega_{Vi}^2$$
$$+ (1 - \gamma\omega_{Vi}/\omega)(\Gamma_j^2 + \Gamma^2) - \gamma\omega_{Vi}\Omega_i^2/\omega = 0. \tag{1.71}$$

Here $\omega_{Vi} = -c\,T_i k_2 \partial_1 V/eV$, $\omega_{Pi} = c\,T_i(k_2\partial_1 P_i - k_1\partial_2 P_i)/e\,P_i$, $\omega_{ni} = cT_i(k_2\partial_1 n_0 - k_1\partial_2 n_0)/en_0$, $\omega_{Ti} = \omega_{Pi} - \omega_{ni}$, and Ω_i^2 is equal to Ω_e^2 with the electron temperature replaced by the ion temperature.

The terms describing drift effects due to density and temperature inhomogeneities complicate the analysis of (1.71). However, the behaviour of the flute instability with $T_e = 0$ coincides qualitatively with the case $T_e \gg T_i$ discussed above. Equation (1.71) has no complex roots when $(|\omega_{Vi}|^2, |\omega_{Pi}|^2) \gtrsim \Gamma_j^2 + \Gamma^2$. This means flute perturbations whose wavelengths are short enough compared to r_s are stable. The root of (1.71) whose imaginary part is positive has the form

$$\omega = \frac{1}{2}\omega_{Pi} - \frac{1}{2}\omega_{Vi}\frac{\Omega_i^2}{\Gamma_j^2 + \Gamma^2} + i(\Gamma_j^2 + \Gamma^2)^{1/2}, \tag{1.72}$$

when $\Gamma_j^2 + \Gamma^2 \gg (|\omega_{Vi}|^2, |\omega_{Pi}|^2)$. From this one can deduce that the real part of the frequency of unstable flute perturbations is controlled by the drift frequency ω_{Pi} and the ion magnetic drift frequency ω_{Vi} to within an order of magnitude.

It is of interest to apply these results to the dipole magnetic field configuration which is frequently regarded as a zeroth approximation to the actual magnetic field of the Earth. The Euler coordinates are given by the following functions:

$$\begin{aligned} x^1 &= k_0 \cos^2\theta/r, \\ x^2 &= \varphi, \\ x^3 &= \sin\theta/r^2, \end{aligned} \tag{1.73}$$

where φ is the eastward geomagnetic longitude, θ the geomagnetic latitude, and r the radial distance (see Akasofu and Chapman, 1972). The Earth's magnetic field is assumed to be formed by a point dipole with moment k_0 which points southward and is located at the coordinate origin. We calculate the following quantities in these coordinates:

$$\begin{aligned} &V = 0.91k_0^3(x^1)^{-4}, && V_{11} = 0.6k_0^5(x^1)^{-8}, \\ &V_{22} = 0.68k_0^5(x^1)^{-6}, && V^{11} = k_0^3L^{-3}(x^1)^{-2}, \\ &V^{22} = L^{-2}k_0^3(x^1)^{-4}(4L-3)^{-1}, && V_{11}' = 0.8(x^1)^4L^5/k_0^5, \\ &V_{22}' = 3.2(x^1)^6L^5/k_o^5. \end{aligned} \tag{1.74}$$

In magnetospheric physics the position of a magnetic field line is usually characterized by a dimensionless parameter L that is related to the coordinate x^1 by means of $x^1 = k_0/LR_E$, where R_E is the Earth's radius. Assuming that plasma pressure in the magnetosphere is distributed according to $p = p_0(L/L_0)^{-n}$, where p_0 is the pressure on the field line with parameter L_0, and using (1.42), we find the growth rate of the ordinary flute instability for the value $\gamma = 5/3$ to be

$$\Gamma^2 = \frac{6k_2^2T(n - 20/3)}{m_iL^2R_E^2(k_2^2 + (k_0^2/L^2R_E^2)k_1^2)}\,. \tag{1.75}$$

This yields the well-known result for a dipole field, that the plasma is unstable with respect to flute perturbations for $n > 20/3$. Formula (1.59) yields a steady-state longitudinal current of

$$j_{\parallel} = -1.8cL^4(4L^2 - 3L)^{1/2}\partial_2 p^0/R_EB_{equ}, \tag{1.76}$$

where $B_{equ} = k_0/R_E^3$ is the magnetic field at the Earth's equator. The growth rate for the flute instability with a variable pressure over the surfaces of equal specific volume (1.64) in a dipole field has the form

$$\Gamma_j^2 = \frac{3.4k_1k_2B_{equ}^2R_E}{cm_in_0L^7(4L^2 - 3L)^{1/2}(k_2^2 + (k_0^2/L^2R_E^2)k_1^2)}j_{\parallel}\,. \tag{1.77}$$

The highest growth rate is attained when $k_2 = \pm\, k_0k_1/LR_E$ and equals

$$\Gamma_j^2 = \frac{1.7B_{equ}}{m_in_0cL^6(4L^2 - 3L)^{1/2}R_E}\,|j_{\parallel}|\,. \tag{1.78}$$

Lastly, we give an expression for the threshold value of the longitudinal current (1.66) and the magnetic drift frequency

$$j_{\parallel}^{*} = \frac{0.63m_in_0cL^6(4L^2 - 3L)^{1/2}R_E}{B_{equ}}\,|\Gamma^2|\,, \tag{1.79}$$

and

$$\omega_V = -\frac{4cTk_zL}{eB_{equ}R_E^2}. \tag{1.80}$$

These formulas can be used to evaluate various quantities that characterize the Earth's magnetosphere.

1.5. ROSSBY WAVES IN THE ATMOSPHERE

Waves in the atmosphere and in the ocean are similar to plasma waves. Thus, Sagdeev (1966) notes a similarity between the ion acoustic and the magnetosonic solitons on the one hand and solitons of long gravity waves on water on the other. The similarity between Rossby waves in the atmosphere and drift waves in plasmas is of special interest. Rossby waves are a continuation of sonic and long gravity waves. When the acoustic wavelength in the atmosphere exceeds its depth, the compressibility of air becomes unimportant. The role of compressibility is now played by changes in the effective depth in the gravitational field. The acoustic waves then transform smoothly into long gravity waves. Particle oscillations in gravity waves occur along the horizontal component of the gradient of the pressure perturbation. Barotropic and internal modes can be distinguished. The phase of the particle oscillations does not depend on height in a barotropic while it is significantly dependent on mode height in an internal mode. Barotropic modes in the atmosphere and oceans are described by the shallow water equations plus the Coriolis force

$$d_t\mathrm{v} = -\nabla_\perp(gH) + \Omega(\mathrm{v}\times\zeta), \tag{1.81}$$

$$\partial_t H + \mathrm{div}\,\mathrm{v}H = 0, \quad d_t \equiv \partial_t + \mathrm{v}\cdot\nabla_\perp, \tag{1.82}$$

where v is the horizontal component of the atmospheric velocity, $\Omega = 2\omega_0 \sin\alpha$ is the Coriolis frequency (parameter), α is the latitude, ω_0 the planet's angular rate of rotation, H the effective depth of the atmosphere, g the acceleration of gravity, and ζ the unit vector along the vertical.

In this model the atmosphere is treated as an incompressible shallow fluid of depth H. We shall investigate the spectrum of small oscillations for the system (1.81) and (1.82). We write $H = H_0(1 + h)$, where H_0 is the unperturbed depth which may depend weakly on the coordinates. Equations (1.81) and (1.82) are linearized with respect to h and v, and a local set of coordinates x, y, z is defined, with the x-axis pointing eastward, the y-axis northward, and the z-axis along the vertical. All perturbed quantities are assumed to vary in space and time as

$$\sim \exp(-i\omega t + i\mathbf{k}_\perp \cdot \mathbf{r})$$

For this approximation to be valid it is necessary that the wavelength in the horizontal direction should be small enough and be in the range $k_\perp H_0 \ll 1 \ll k_\perp R$, where R is the planet's radius (typical size of the inhomogeneity).

From (1.81) and (1.82) we then obtain

$$\omega(1 + k_\perp^2 r_R^2 - \omega^2/\Omega^2) = k_x v_R, \tag{1.83}$$

where $r_R = c_g/\Omega$ is the Rossby radius (deformation radius), $c_g = (gH_0)^{1/2}$, $v_R = 0.5H_0\omega_0(1 + \varkappa \sin\alpha)\cot\alpha$, and $\varkappa = d \ln H_0/d \sin\alpha$ is the Rossby velocity (drift velocity).

Equation (1.83) describes two branches separated by a gap (Figure 1.2). The branch whose frequencies are high compared to Ω corresponds to long gravity waves on shallow water

$$\omega^2 = k_\perp^2 c_g^2. \tag{1.84}$$

As the frequency decreases, the Coriolis force grows in importance. It acts to shift the direction of the particle oscillations from the direction of the pressure gradient. When $\omega \sim \Omega$, (1.83) yields the dispersion relation for inertial (gyroscopic) waves in a shallow rotating atmosphere

$$\omega = \Omega(1 + k_\perp^2 r_R^2/2). \tag{1.85}$$

These frequencies are typical of typhoons in the tropical latitudes. They are analogous to cyclotron waves in plasmas (Petviashvili and Pokhotelov, 1983).

Finally, in the limit of low frequencies compared to Ω, the particles oscillate almost perpendicular to the wave vector. The rotational part of the particle velocity is large compared with the potential part. From (1.83) we get

$$\omega = \frac{k_x v_R}{1 + k_\perp^2 r_R^2}. \tag{1.86}$$

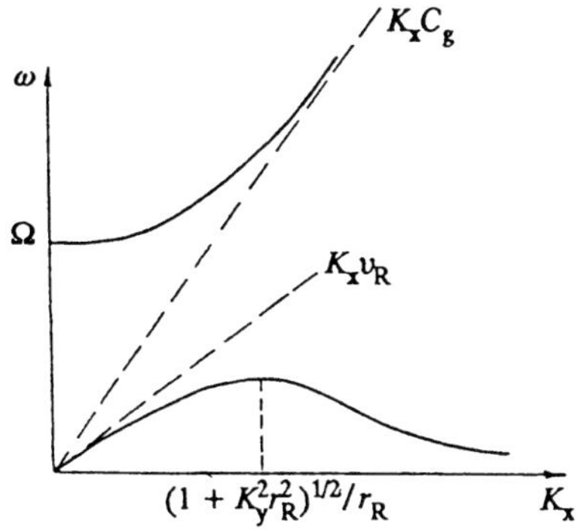

Figure 1.2. Spectrum of waves in shallow rotating water. The upper branch corresponds to inertial waves and is separated from the lower branch (Rossby waves) by a wide gap. When k is large, the inertial waves become long gravity waves travelling at velocity c_g.

The waves described by this equation bear the name of the Swedish geophysicist Rossby who was the first to appreciate their role in weather formation in the atmosphere and oceans. In recent years, interest in these waves has been revived in connection with attempts to model various physical phenomena, including the Gulf Stream rings (Flierl, 1979), atmospheric blocking (Baines, 1983), and the Great Red Spot of Jupiter (Petviashvili, 1980).

The phase of the oscillations in the internal modes depends on the altitude. For this reason, one should take into account the inhomogeneity of the atmosphere along the vertical and the vertical component of the velocity. These are described by the more complicated equations:

$$d_t \mathbf{v}_\perp = -\nabla p/\rho + \Omega(\mathbf{v}_\perp \times \zeta); \tag{1.87}$$

$$\partial_z p = -\rho g; \tag{1.88}$$

$$d_t p = -\gamma p \operatorname{div} \mathbf{v}; \tag{1.89}$$

and

$$\partial_t \rho + \operatorname{div} \rho \mathbf{v} = 0, \tag{1.90}$$

where $\perp$ denotes a horizontal component as before. In the low-frequency approximation adopted here, the pressure is determined in the hydrostatic approximation (1.88). Pressure oscillations are assumed to be adiabatic. This is true when the period of the oscillations, T, is small compared to the typical time over which the temperature becomes uniform. The conditions $\Omega T \gtrsim 1$ is also necessary.

As in the case of the barotropic modes, Equations (1.87)–(1.90), yield a dispersion relation for the internal modes in the quasiclassical approximation:

$$\omega = \frac{k_x v_R}{k_\perp^2 r_R^2 + k_z^2 r_i^2}, \tag{1.91}$$

where k_z is the vertical wavenumber, which takes on discrete values to be determined from the boundary conditions, $r_i \equiv r_R \Omega / N$ is the so-called internal Rossby radius, and $N^2 = g/H_0 - g^2/c_s^2$ is the square of the Brunt–Väsälä frequency.

In the simplest case we have $k_z = \pi m / 2H_0$, where m is an integer.

It can be seen from (1.91) that the phase velocities of the internal modes along the horizontal are small compared to those for the barotropic modes and also have a westward direction. For the terrestrial atmosphere $N \simeq 10^{-2}\,\mathrm{s}^{-1}$, and $\Omega \simeq 10^{-4}\,\mathrm{s}^{-1}$, so that $\Omega/N \ll 1$. The Rossby radius is large, $r_R \simeq 2000$ km, while the internal Rossby radius is 30 km; hence the internal modes are easily accomodated by the terrestrial atmosphere and oceans.

2. Nonlinear equations of the acoustic type

2.1. MODEL EQUATIONS FOR ACOUSTIC WAVES IN ISOTROPIC MEDIA

The modern theory of nonlinear waves is based on model equations. The best known among these is the Korteweg–de Vries equation, which was derived to describe nonlinear waves on shallow water.

Subsequent developments have shown that this equation is a rather general model for 1D nonlinear waves of the acoustic type propagating in a variety of dispersive media, such as plasmas or solids. It has been shown that many types of oscillations, for example ion acoustic and magnetosonic waves in plasmas, phonons in solids and condensed helium, and magnetoelastic waves in antiferromagnets, obey the Korteweg–de Vries equation.

The basic idea in deriving model equations of this sort is the scheme of Korteweg and de Vries, who simplified general wave equations by expanding them in a perturbation theory expansion with the nonlinear parameters and dispersion parameters being treated as small quantities of the same order of magnitude. As a result of this idea, such concepts as soliton, self-focusing, collapse, the number of quasi-particles, the total wave energy, etc., have gained universal applicability. This double expansion method was not used for a long time, because the simplified equations obtained from it were unusual and apparently hard to analyse.

The interest in the Korteweg–de Vries equation was revived by Zabusky and Kruskal (1965), who discovered the unusual properties inherent in solutions to the Korteweg–de Vries equation. They demonstrated numerically that solutions of this equation contain solitary waves that do not interact with one another; these have been termed solitons because of their similarity to particles. Gardner *et al.* (1967) showed that the Korteweg–de Vries equation can be handled by solving a set of linear equations. The method was developed by Lax (1968) and

Zakharov *et al.* (1980) and has come to be known as the inverse scattering problem (ISP) technique.

In the early 1970s attempts have been made to get out of a 1D approximation. Kadomtsev and Petviashvili (1970) used the idea of the parabolic Leontovich equation and the Korteweg–de Vries equation to derive an equation describing the propagation of weakly non-1D acoustic waves in a dispersive medium. That equation has the same degree of universal applicability as the Korteweg–de Vries equation. We give a derivation of that equation using ion acoustic waves in an isotropic plasma as an example.

Plasmas can be regarded as isotropic if the influence of the magnetic field is small. The only acoustic mode that is possible under these conditions is ion acoustic waves in the frequency range $\omega_{Bi} \ll \omega < \omega_{Pi}$. Such waves with infinitesimal amplitude were discussed in chapter 1. Here we shall examine finite-amplitude effects. Sagdeev (1966) was the first to investigate nonlinearity in ion acoustic waves. In these oscillations the electrons are distributed according to a Boltzmann distribution

$$n_e = n_0 \exp(\Phi), \quad \Phi \equiv e\varphi/T_e \tag{2.1}$$

while the equations of motion and continuity for the ions have the forms

$$d_t\mathbf{v} = -c_s^2\nabla\Phi + \omega_{Bi}(\mathbf{e}_z \times \mathbf{v}), \tag{2.2}$$

and

$$\partial_t n_i + \operatorname{div} n_i\mathbf{v} = 0, \quad d_t \equiv \partial_t + \mathbf{v}\cdot\nabla, \tag{2.3}$$

where $\mathbf{v}$ is the hydrodynamic velocity of the ions. The ion pressure is neglected, as before, because T_i is low. The departure from charge neutrality, which is significant for small wavelengths, is described by Poisson's equation

$$r_D^2\nabla^2\Phi = (n_e - n_i)/n_0. \tag{2.4}$$

As in chapter 1, we shall first consider the effect of nonlinearity on the propagation of high frequency ion sound waves with frequencies that are high compared to ω_{Bi}. The last term in (2.2) can then be discarded and the plasma becomes isotropic. Suppose we have a packet whose typical dimension in the z-direction is small compared with those in the other directions. In this approximation the effects of dispersion and diffraction on the wave packet can be neglected when deriving nonlinear corrections in simplified form. It is enough to include the finiteness of just the z-component of the ion speed, while the other components can be disregarded. Equations (2.2) and (2.3) then yield

$$\partial_t v + v\partial_z v = -c_s^2 \partial_z \Phi, \quad v \equiv v_z, \tag{2.5}$$

$$\partial_t \ln n_i + v\partial_z \ln n_i + \partial_z v = 0. \tag{2.6}$$

The left-hand side of (2.4) describes the dispersion of the ion sound which we treat as small. For this reason the nonlinear corrections can be obtained under the assumption of quasi-neutrality

$$\ln n_i = \ln n_e = \Phi, \tag{2.7}$$

where (2.1) has been utilized. We seek v and Φ as functions of the form

$$v = v(\zeta, \tau, x, y), \quad \zeta \equiv z - c_s t, \quad \tau = t. \tag{2.8}$$

A parabolic approximation will be employed below, assuming the dependence on the first argument is much stronger than that on the others.

Equations (2.5) and (2.6) can be rewritten in these variables as follows:

$$\partial_\tau v + v\partial_\zeta v = c_s \partial_\zeta (v - c_s \Phi); \tag{2.9}$$

$$\partial_\tau \Phi + v\partial_\zeta \Phi = -\partial_\zeta (v - \tau_s \Phi). \tag{2.10}$$

In the zeroth approximation, given that the amplitude is small, i.e. that $\Phi \ll 1$, we have $v \simeq c_s \Phi$. Substituting this relation into the small terms in (2.9) and (2.10), we get

$$\partial_\tau \Phi + c_s \Phi \partial_\zeta \Phi = 0. \tag{2.11}$$

This equation only includes the nonlinearity, but the dispersion and diffraction terms which generally are of the same order as the nonlinear ones, have been omitted. These terms were included independently in the linear approximation in equation (1.14). Combining (1.14) and (2.11), we obtain a single equation:

$$\partial_\zeta(\partial_\tau\Phi + c_s\Phi\partial_\zeta\Phi - (c_s r_D^2/2)\partial_\zeta^3\Phi) = -(c_s/2)\nabla_\perp^2\Phi. \qquad (2.12)$$

Its dimensionless form is

$$\partial_z(\partial_t\Phi + \Phi\partial_z\Phi + \partial_z^3\Phi) = \sigma\nabla_\perp^2\Phi. \qquad (2.13)$$

This is known as the Kadomtsev–Petviashvili equation. Here $\sigma = \pm 1$ is the sign of the dispersion. A broad class of acoustic wave equations, both in isotropic and in many anisotropic media, can be reduced to this equation. We have $\sigma = -1$ corresponding to media with negative dispersion in the case of ion sound. Positive dispersion is typical of capillary waves on the surface of a fluid and, under certain conditions, of phonons in condensed helium. One example of such waves in a cold plasma with $\beta \ll 1$ is the fast magnetosonic mode with frequencies that are low compared to the cyclotron frequency when propagation takes place obliquely to the magnetic field. When the dependence on x, y can be disregarded, that is, when we are concerned with a one-dimensional packet, equation (2.13) reduces to the Korteweg–de Vries equation.

The sign of the dispersion is not essential in the one-dimensional case, because it can be changed by a similarity transformation. In (2.13), however, the sign of the dispersion separates the equation into two classes which cannot be transformed into one another.

2.2. 1D AND 2D SOLITON SOLUTIONS TO THE KADOMTSEV–PETVIASHVILI EQUATION AND THEIR STABILITY

Equation (2.13) has a solution in the form of 1D solitons:

$$\Phi_0 = \frac{3u}{\mathrm{ch}^2(\zeta/\sqrt{2u})}, \qquad (2.14)$$

where $u > 0$ is the speed and $\zeta \equiv z - ut$. The one-dimensional solutions of a general form that depend only on z and t coincide with the solutions to the Korteweg–de Vries equation have been found by the ISP method (Lax, 1968). Zakharov *et al.* (1980) and Dryuma (1974) have pointed out that 2D solutions to the Kadomtsev–Petviashvili equation can be obtained in analytical form. 2D soliton solutions to the Kadomtsev–Petviashvili equation were found numerically by Petviashvili (1976) and in a general form by Bordag *et al.* (1977) using a new modification of the ISP method. In a particular case the solution has the form of two 1D non-interacting solitons. Soliton solutions to the Korteweg–de Vries equation are stable for either sign of the dispersion. Kadomtsev and Petviashvili (1970) have investigated these solutions for stability in a 2D space, when they are described by the Kadomtsev–Petviashvili equation. To do this, the non-dimensionalized equation (2.13) is linearized with respect to small perturbations φ of the exact solution (2.14). This leads to the equation

$$\partial_z\left[\partial_t\varphi + \partial_z(\varphi\Phi_0) + \partial_z^3\varphi\right] = \sigma\nabla_\perp^2\varphi \tag{2.15}$$

Giving the expression for Φ_0, we seek a solution to (2.15) in the form

$$\varphi = \psi(\zeta)\exp(-i\omega t + ikx), \quad \zeta \equiv z - ut. \tag{2.16}$$

Substituting (2.16) into (2.15) yields an equation in ψ:

$$L\psi = i\omega\partial_\zeta\psi - \sigma k^2\psi, \tag{2.17}$$

where the operator L is defined as

$$L = \partial_\zeta^2(-u + \Phi_0 + \partial_\zeta^2). \tag{2.18}$$

Here we assume that $k^2 \simeq \omega^2 \ll \omega$. We have, thereby, restricted ourselves to considering long-wavelength low-frequency perturbations. In that case, ψ can be represented as a power series in the small parameter ω:

$$\psi = \psi_0 + \psi_1 + \psi_2 + \dots$$

Substituting this expansion into (2.17) and equating terms of the same degree, we obtain the following sequence of equations

$$\begin{aligned} L\psi_0 &= 0, \\ L\psi_1 &= i\omega\partial_\zeta\psi_0, \\ L\psi_2 &= i\omega\partial_\zeta\psi_1 - \sigma k^2\psi_0. \end{aligned} \tag{2.19}$$

The solution of the first equation has the form

$$\psi_0 = \partial_\zeta\Phi_0. \tag{2.20}$$

Substituting this solution into the second of equations (2.19) and solving it, we get

$$\psi_1 = i\omega\partial_u\Phi_0. \tag{2.21}$$

Lastly, with the aid of (2.21), from (2.19) we obtain

$$L\psi_2 = -\sigma k^2\partial_\zeta\Phi_0 - \omega^2\partial_u\partial_\zeta\Phi_0. \tag{2.22}$$

This equation is inhomogeneous and, in general, does not have finite solutions for arbitrary terms on the right-hand side. For a solution to be finite, the right-hand side of (2.22) must, as is well known, be orthogonal to solutions of the equation which is conjugate to (2.22), but without the right-hand side. Multiplying (2.22) by Φ_0 and integrating over ζ, we find that the left-hand side vanishes, while the right-hand one yields a dispersion relation for small soliton oscillations

$$\omega^2 = -\sigma k^2\partial_u(\ln\textstyle\int\Phi_0^2 d\zeta). \tag{2.23}$$

Using the explicit expression for the soliton solution (2.14), we obtain

$$\omega^2 = -3\sigma k^2/u. \tag{2.24}$$

One can see from (2.24) that the 1D soliton is stable when $\sigma = -1$ and unstable when $\sigma = +1$. Solitons are thus stable in media with negative dispersion and unstable in media with positive dispersion. The method used here to investigate stability is not quite rigorous, since the third approximation increases in space as we move away from the soliton. A rigorous examination of stability can be performed by the ISP method (Zakharov *et al.*, 1980). This confirms the present results and demonstrates that instability occurs only for wavelengths greater than the soliton width. This suggests that stable 2D solitons must be possible in two dimensions in media with positive dispersion. Such soliton solutions to the Kadomtsev–Petviashvili equation have been found numerically by Petviashvili (1976a), and afterwards in analytical form by Bordag *et al.* (1977). The latter reference also contains solutions in the form of a set of an arbitrary number of different solitons. If the solution to (2.13) is sought in the form of a soliton

$$\Phi_0 = u\psi(\zeta,\eta),\quad \zeta \equiv (z-ut)\sqrt{u},\quad \eta \equiv ux, \tag{2.25}$$

then, after substituting (2.25) into (2.13), we obtain an equation for a 2D soliton

$$\sigma\partial_\eta^2\psi + \partial_\zeta^2\psi = \partial_\zeta^4\psi + (1/2)\partial_\zeta^2\psi^2. \tag{2.26}$$

Equation (2.26) is the equation for a nonlinear string with dispersion in which the time is represented by $t = i\eta\sqrt{\sigma}$. The time is purely imaginary in the case $\sigma = 1$. It is readily verified by direct substitution that when $\sigma = 1$, (2.26) has a 2D soliton solution of the form

$$\psi = \frac{24(\eta^2 - \zeta^2 + 3)}{(\eta^2 + \zeta^2 + 3)^2}. \tag{2.27}$$

Zaitsev (1983) has obtained a solution of (2.26) for $\sigma = 1$ in the form of an N-soliton group using the Hirota substitution (Hirota, 1973)

$$\psi = 12\partial_\zeta^2 \ln \det M, \tag{2.28}$$

where

$$M_{ij} = \delta_{ij} \pm m_{ij} \exp(\theta_i + \theta_j). \tag{2.29}$$

Here δ_{ij} is the Kronecker delta, θ_i is the phase of the i-th soliton, i and j take on values from 1 to N, with $2\theta_i = k_i(\zeta + \varkappa_i\eta)$; $\varkappa_i = k_i(k_i^2 - 1)$; and $m_{ij}^2 = -\sigma k_i k_j / [-7k_i k_j + 2(k_i - k_2)^2 + ((k_i^2 - 1)(k_j^2 - 1))^{1/2} + 1]$.

Equation (2.29) can yield a great variety of steady-state multisoliton solutions to the 2D Kadomtsev–Petviashvili equation. In particular, (2.27) is obtained for $N = 2$ in the limit $k_1 \to 0$, $k_2 \to 0$, while $\varkappa_1 \to -\varkappa_2 \to i$. We then have $\det M \to \eta^2 + \zeta^2 + 3$. Such solutions are not stable in 2D space, but an individual soliton of the form (2.27) is stable. We now examine the stability of this solution in 3D space. Two possibilities exist: the diffraction term from the third coordinate y has the same sign in an isotropic medium as that from x, while in anisotropic media (for instance, in antiferromagnets) the sign may be opposite. In the former case, instability is proved in a similar fashion to the case of 1D soliton. In the second case, however, soliton instability is not established by proceeding according to the same scheme. Nevertheless, this does not prove that the soliton is stable, since this method does not cover all kinds of possible soliton perturbations. Therefore, we shall consider this case in accordance with the scheme proposed by Kuznetsov and Turitsyn (1982) and Kusmartsev (1984). We rewrite (2.23) in a form that explicitly includes the anisotropy owing to diffraction,

$$\partial_z(\partial_t\Phi + \Phi\partial_z\Phi + \partial_z^3\Phi) = \partial_x^2\Phi - \partial_y^2\Phi. \tag{2.30}$$

When $\partial_y\Phi = 0$, (2.30) has a solution in the form of a 2D soliton described by (2.26) for $\sigma = 1$. Consider the equation for small disturbances φ of this solution

$$\partial_z(\partial_t\varphi + \partial_z(\varphi\Phi_0) + \partial_z^3\varphi) = \partial_x^2\varphi - \partial_y^2\varphi, \tag{2.31}$$

where Φ_0 is given by (2.27).

We seek a solution of the form

$$\varphi = \psi(\zeta, x)\exp(-i\omega t + iky), \quad \zeta \equiv z - ut. \tag{2.32}$$

Substituting (2.32) into (2.31) yields an equation for $\psi(\zeta, x)$ of the form (2.17), where however the operator L is two-dimensional and is given by

$$L = \partial_\zeta^2(-u + \Phi_0 + \partial_\zeta^2) - \partial_x^2. \tag{2.33}$$

As before, we assume $\omega \sim k$ to be small and seek solutions in the form of the series (2.16) in powers of a small parameter. We then obtain a sequence of equations, (2.17) through (2.19), with the operator (2.33). In contrast to the 1D case, we have two localized solutions in the zeroth approximation: (2.20) and

$$\psi_0 = \partial_x\Phi_0. \tag{2.34}$$

If the solution in the zeroth approximation is taken in the form (2.20), it turns out that the spectrum of this mode is real. This means that the soliton is stable with respect to perturbations that are antisymmetric in ζ and symmetric in x. Now consider perturbations with the other symmetry, (2.34). Substitution of (2.34) into (2.19) gives

$$\psi_1 = -\frac{i\omega}{2}x\partial_\zeta\Phi_0. \tag{2.35}$$

Using (2.35) we obtain, from (2.19)

$$L\psi_2 = (\omega^2 x/2)\partial_\zeta^2\Phi_0 + k^2\partial_x\Phi_0. \tag{2.36}$$

As in the case of a 1D soliton, the finiteness condition for ψ_2 yields a dispersion relation describing the spectrum of small oscillations for a 2D soliton in an anisotropic medium,

$$\omega^2 = -4k^2 \frac{\int (\partial_x \Phi_0)^2 d_\zeta dx}{\int (\partial_\zeta \Phi_0)^2 d_\zeta dx}. \tag{2.37}$$

It follows from (2.37) that a 2D soliton of the Kadomtsev–Petviashvili equation is unstable in 3D space, whatever the sign of the dispersion relative to the third coordinate.

In contrast to the results of the preceding discussion, where the instability of a 1D soliton in 2D space was found to be controlled by the sign of the dispersion, here the instability occurs in 3D space for either sign of the dispersion. This is associated with the fact that the small free oscillations of a 2D soliton have two eigen modes with different symmetries.

2.3. STABILITY OF 3D KADOMTSEV–PETVIASHVILI SOLITONS. COLLAPSE AND SELF-FOCUSING

Petviashvili (1976a) has found that the Kadomtsev–Petviashvili equation in positively dispersive media has 2D and 3D soliton solutions. Since it has been shown that 2D solutions are unstable in 3D space, an investigation of the stability of the 3D solutions is of interest. Note that the following integrals of motion are conserved in the 3D Kadomtsev–Petviashvili equation:

$$P = \int \Phi^2 d^3x \tag{2.38}$$

and

$$H = \int \left[(\partial_\zeta \Phi)^2 + \sigma(\nabla_\perp \mu)^2 - (2/3)\Phi^3 \right] d^3x, \tag{2.39}$$

where $\partial_s \mu = \Phi$ and P represents the momentum of a wave packet. One can use the integrals (2.38) and (2.39) to write the Kadomtsev–Petviashvili equation in the Hamiltonian form

$$\partial_t \Phi = \partial_z \frac{\delta H}{\delta \Phi}. \tag{2.40}$$

Thus, H can be referred to as the Hamiltonian for the Kadomtsev–Petviashvili equation.

3D solitons might be expected to be stable, if H were bounded from below while P constant. We shall show that this is not the case. Suppose that a 3D soliton solution to (2.13) has the form $\Phi = f(\zeta, r)$, where ζ and r are cylindrical coordinates. We now introduce a new function

$$g = (1+\xi)f\left[(1+\eta)^2\zeta,\ (1+\xi)(1+\eta)r\right]. \tag{2.41}$$

Here ξ, η are some small numbers that model the extent of the soliton and are such that the momentum is unchanged when g is substituted in P. Substituting (2.41) into the Hamiltonian H, we see that it is a function of ξ and η of the form

$$H \simeq H_0 + A_1\xi + A_2\eta + B_{11}\xi^2 + 2B_{12}\xi\eta + B_{22}\eta^2 + \ldots \tag{2.42}$$

If f is a soliton solution, it yields an extremum of H for fixed P. This can be seen from the fact that the soliton equation (2.13) is identical with Euler's equation for the functional $L = H + \lambda P$. At the extremum of H, we must have $A_1 = A_2 = 0$. Direct calculation shows that

$$B_{11} = 10I,\ \ B_{12} = -8I,\ \ B_{22} = 2I, \tag{2.43}$$

where $I = \int (\partial_r f)^2 d^3x$.

The quadratic form (2.42) is thus, in view of (2.43), not positive definite. The soliton solution is, therefore, a saddle point in the function space. If H and P are assumed to be the only integrals of the 3D Kadomtsev–Petviashvili equation, then it follows that the 3D soliton solutions to the Kadomtsev–Petviashvili equation are not stable. The method used here to examine the instability of the soliton is due to Derrick (1964). As the Hamiltonian is not bounded below in 3D space for fixed P, which follows from (2.42), it can take on any values, including negative ones. As Kuznetsov *et al.* (1983) have shown, a wave packet tends to spread when $H > H_c$ (where H_c is the value of the Hamiltonian for the 3D soliton solution), while when $H < H_c$, self-compression, accompanied by an infinite growth of amplitude, can occur (for H and P fixed). This phenomenon is called collapse. It can only occur with positive dispersion. This has been demonstrated by numerical calculations in Kuznetsov *et al.* (1983). The results of the calculation are shown in Figure 2.1. The central portion of the wave packet is seen to lag

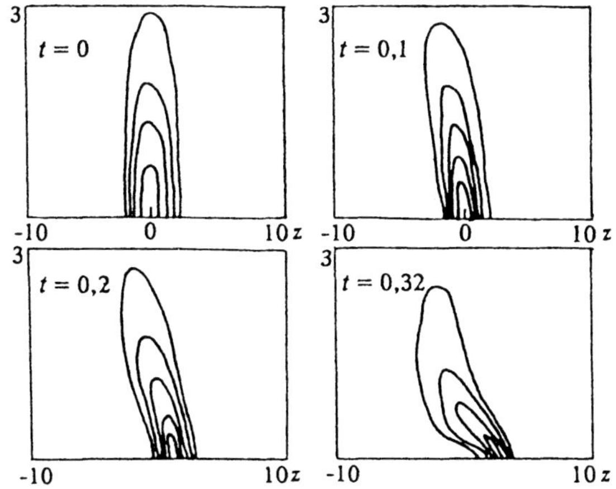

Figure 2.1. Contours of $\Phi(z, r_\perp)$ at consecutive instants of time. The wave packet amplitude increases by a factor of 7 over a time $t = 0.32$ from 4 at $t = 0$.

behind the edge. Intensive radiation is observed from the cavity and this causes a reduction in H and P. In principle, this radiation could halt the collapse. It turns out, however, that H decreases more rapidly than P during the radiation. For this reason the radiation enhances the collapse.

When a beam of sound propagates in a medium with positive dispersion, another interesting effect can be observed: self-focusing, which can also be described by the Kadomtsev–Petviashvili equation. Nonlinear self-compression dominates diffractive spreading of the beam during self-focusing. As the cross-sectional area of the beam decreases, the beam intensity grows to quite high levels. It has been pointed out by Ozhogin *et al.* (1983), and Turitsyn and Falkovitch (1985), that self-focusing of sound can occur in solids with especially strong nonlinearity. These include antiferromagnetic materials with a strong magnetostrictive interaction, that leads to a large nonlinearity in the elastic subsystem of the crystal (Ozhogin and Preobrazhensky, 1977). In most other solids, however, dissipation becomes stronger than elasto-nonlinear effects as the amplitude of the sound increases. Experimental estimates of the nonlinearity coefficients in antiferromagnetic materials can be found in Ozhogin *et al.* (1983), Ozhogin and Preobrazhensky (1977) and elsewhere.

Self-focusing can also occur in plasmas when a beam of magnetosonic waves propagates at an angle to the magnetic field that is not near either zero or $\pi/2$. To demonstrate this, one can use the Kadomtsev–

Petviashvili equation in a form that is suitable for solving initial-value problems. Self-focusing is investigated by deriving a simplified equation suitable for boundary-value problems.

We assume the amplitude $\Phi = \Phi_0(t, x, y)$ to be fixed at the boundary $z = 0$ as a function of time and x and y. In the absence of dispersion and diffraction, this kind of perturbation could propagate in the z direction in accordance with the formula

$$\Phi = \Phi_0(\tau, x, y), \quad \tau \equiv t - z/c_A. \tag{2.44}$$

We shall include the nonlinearity, dispersion, and diffraction in this expression as small corrections. Proceeding as in the derivation of the Kadomtsev–Petviashvili equation for initial-value problems, we introduce a weak dependence of (2.44) on z. We then find that Φ obeys the equation

$$\partial_\tau(\partial_z\Phi + \Phi\partial_\tau\Phi + \partial_\tau^3\Phi) = \sigma\nabla_\perp^3\Phi. \tag{2.45}$$

This equation is identical to the Kadomtsev–Petviashvili equation when the arguments are interchanged and describes wave propagation to the right of the $z = 0$ plane where the corresponding boundary condition is imposed. Self-focusing and collapse both occur only when $\sigma > 0$. We shall make some remarks on the difference between self-focusing and collapse. In the former case, one imposes boundary conditions that are localized in x and y and periodic in time, while in the latter, initial conditions that are localized in all coordinates are imposed. For this reason, self-focusing may occur in a regime where a wave beam undergoes steady-state contraction. However, because of a possible instability, this steady-state regime may change to a fluctuating regime with random parameter distributions.

The problem can be further simplified in the steady state by passing to Whitham's truncated equations (Whitham, 1974); this has been done by Manin and Petviashvili (1984). Figure 2.2 illustrates self-focusing with increasing distance from the source within the Kadomtsev–Petviashvili equation framework in the Whitham approximation. Unfortunately, Manin and Petviashvili (1984) mistakenly assumed magnetic sound has positive dispersion when it propagates perpendicular to a magnetic field. As noted by McMahon (1968), however, the dispersion of magnetic sound propagating across a magnetic field in the range of angles $\alpha < \beta^{1/2}$ is negative; hence, self-focusing does not occur.

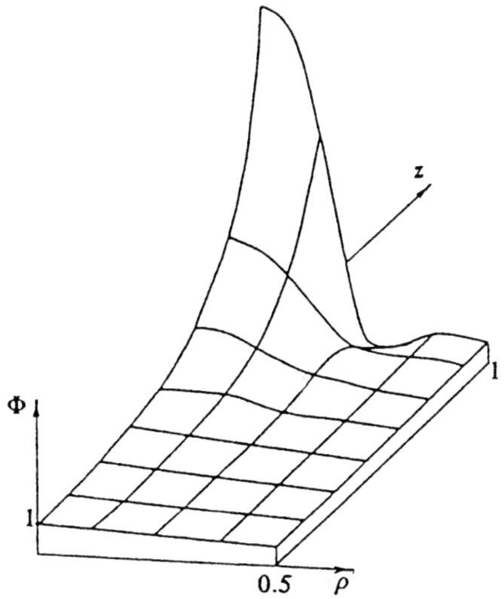

Figure 2.2. Contours of a self-focusing beam of acoustic waves described by (2.45). The wave amplitude is assigned in the form $\Phi = \exp(-\rho^2)$ at the boundary $z = 0$. The values are in dimensionless units.

2.4. THE STABILIZING-MULTIPLIER METHOD FOR SOLITON SOLUTIONS

The simplified equations for the steadystate yield partial differential equations in which the nonlinearity usually enters in a simple fashion. They cannot generally be solved analytically other than in the 1D case. In most cases, however, it is easy to determine whether an equation has a soliton solution. There is no general mathematical theory of these questions for non-1D problems when they are not integrable. Petviashvili (1976a) put forward a simple algorithm that can yield a soliton solution. This algorithm readily yields such a solution by numerical methods. It cannot, unfortunately, always be used to determine whether the localization of the solution is exponential or follows a power-law. Petviashvili (1976a) has pointed out that soliton equations can be reduced to a homogeneous nonlinear integral equation of the following form by using the Fourier or Fourier–Bessel transforms:

$$f = I[f, \lambda]. \tag{2.46}$$

Here I is a nonlinear integral operator acting on f and λ is an eigenvalue of the problem, to be found from the requirement that (2.46) must have a sufficiently smooth solution that vanishes at infinity.

Similarity transformations can be used to eliminate λ in conservative systems (i.e. those without dissipation) with a single definite degree of nonlinearity. If, however, we are concerned with an active medium in which a soliton is maintained by an instability, then λ cannot be eliminated, and it is an important parameter of the problem. Such solitons have been examined by Petviashvili and Tsvelodub (1978) (see also § 3.4, where Langmuir cavitons are discussed). Here, we consider a soliton equation derived from the 2D Kadomtsev–Petviashvili equation. The similarity transformation converts it to the form

$$(\partial_\zeta^2 + \partial_\eta^2 - \sigma\partial_\zeta^4)\Phi = \partial_\zeta^2\Phi^2, \quad \sigma = \pm 1. \tag{2.47}$$

We represent Φ as a Fourier integral:

$$\Phi = \int f_k \exp(ikr)dk. \tag{2.48}$$

(2.47) then reduces to the form

$$\begin{aligned} f_k &= G_k N_k, \quad G_k = k_\zeta^2/(k^2 + \sigma k_\zeta^4), \\ N_k &= \int f_{k'} f_{k-k'}\, dk' \end{aligned} \tag{2.49}$$

It has been noted above that the presence of a soliton solution can be identified by some simple criteria. Here this is evident from the expression for G_k in (2.49). For (2.48) to be solitary, it is necessary that f_k be solitary. But (2.49) has a smooth solitary solution only when the kernel G_k is smooth and solitary; this is true only when $\sigma = 1$, that is, for positive dispersion. At first glance its seems that (2.49) can be solved by iterations. But the operator on the right-hand side of (2.49) is non-compressive, that is, the sequence of iterative solutions diverges. Petviashvili (1976a) has developed the stabilizing-multiplier method which has been used to find a 2D soliton solution to (2.47). This method is a modified iterative procedure and consists in the following: instead of (2.49), we solve the equation

$$\begin{aligned} f_k &= sG_k N_k \equiv K[f], \\ s &\equiv \left[\int f_k^2 dk \Big/ \int f_k G_k N_k dk\right]^n, \end{aligned} \tag{2.50}$$

where K is an integral operator in $\mathbf{k}$ space. The stabilizing multiplier s effectively changes the degree of nonlinearity. It has been found experimentally (Petviashvili, 1976a) that the best convergence is achieved when the degree of nonlinearity equals zero, that is, when $n = 2$ in our case. Equation (2.50) gives a rapidly converging sequence of iterations through the formula

$$f_{i+1} = K\,[f_i], \quad i = 0, 1, 2, \ldots,$$
$$\|f_{i+1} - f_i\| \to 0, \quad \|f\|^2 \equiv \int f_{\mathbf{k}}^2\, d\mathbf{k}. \tag{2.51}$$

In this process, the stabilizing multiplier s approaches 1 as the number of iterations increases.

If the soliton equation is reducible to an ordinary second-order differential equation, this method provides poor accuracy compared with the other procedures. If, on the other hand, the nonlinear equation contains higher-order derivatives or contains partial derivatives, it is best to solve the equation by this method. An analytical solution of (2.47) has been found by Bordag *et al.* (1977). (See § 2.2.) A comparison of the numerical results obtained by the stabilizing-multiplier method (Figure 2.4) with that solution reveals good agreement between the two; hence, we are justified in using this method to study other, more complicated, cases. These cases will be examined in the next section.

2.5. COUPLED SOLITON STATES (MULTI-SOLITONS)

N-soliton solutions to the Kadomtsev–Petviashvili equation were considered in § 2.2. Similar solutions also exist for a number of other model equations. They describe a gas made up of solitons which approach and move apart from one another with no change of shape. Zaitsev (1983) has found solutions to the 2D Kadomtsev–Petviashvili equation in which solitons form coupled states in the shape of an infinite chain or a periodic 2D lattice.

It would be of great interest to find out whether coupled states of a finite number of solitons that move as a whole can exist. In such states, the solitons would be held by the field of their neighbours, like atoms in a molecule. No analytical solutions of this sort for the Kadomtsev–Petviashvili equation have been found. Abramyan and Stepanyants

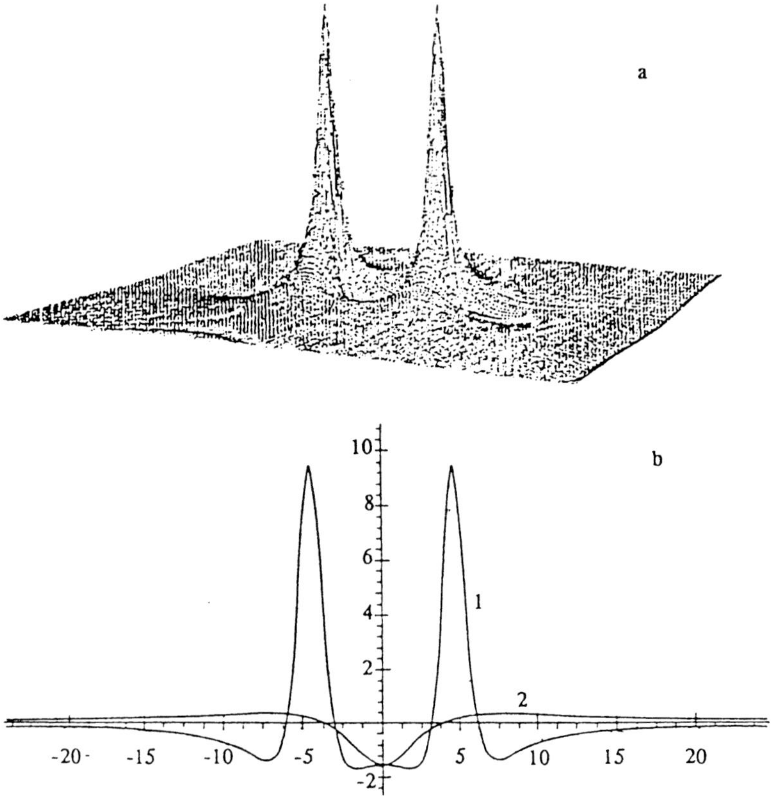

Figure 2.3 (a) contours of the 2D bisoliton solution of (2.26); (b) cross-sections of the bisoliton along (line 1) and perpendicular to (line 2) the direction of motion.

(1985) used the stabilizing-multiplier method to find a coupled state of two solitons (bisoliton). Figure 2.3 shows a contour plot of this solution. For purposes of comparison, Figure 2.4 shows a solution in the form of a single soliton that travels at the velocity of the bisoliton whose analytical expression is given by (2.27). A comparison of the two plots shows that a bisoliton forms when one soliton is trapped in the potential well from another soliton with the same parameters.

Corrections to these equations can produce qualitatively new effects, so they should sometimes be included, even if they are small. These include the corrections to the dispersion. If a correction term weakens dispersion, then the soliton may become unstable or radiate. The latter

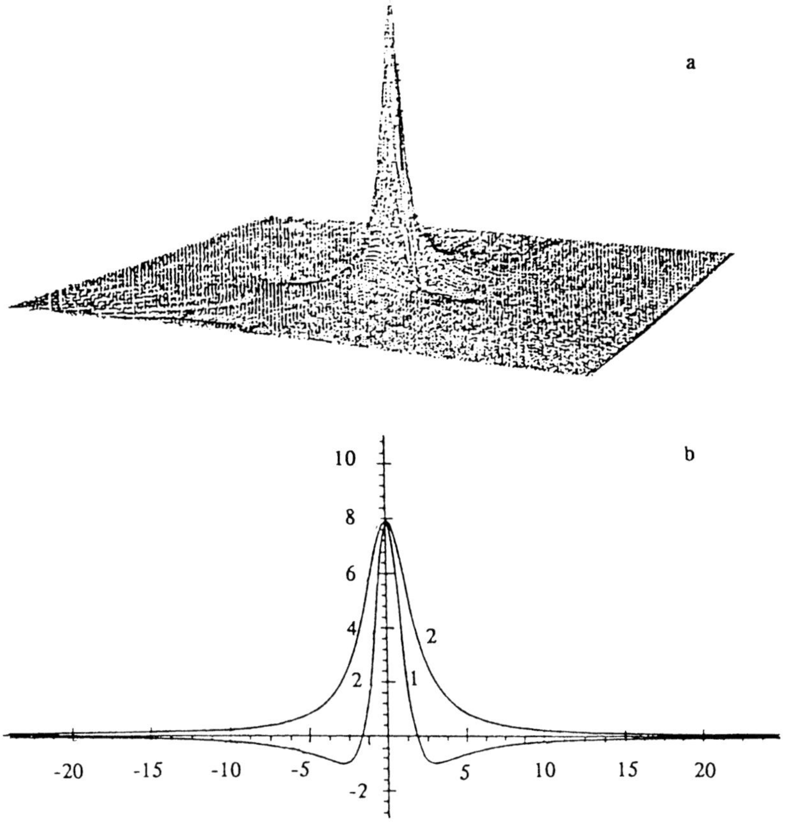

Figure 2.4 (a) contours of the 2D soliton solution of (2.26); (b) cross-sections of the soliton along (line 1) and perpendicular to (line 2) the direction of motion.

means that the equation with the correction may have no strictly steady-state solitary solutions. If a correction makes the dispersion stronger, then multisoliton solutions may appear.

A peculiar situation arises in the case of surface waves in shallow water. Abramyan and Stepanyants (1985) have shown that such waves obey the following equation in a reference frame travelling at the velocity $C_g = (gH_0)^{1/2}$ (g is the acceleration of gravity, H_0 is the unpertubed depth of the water):

$$\partial_z(\partial_t h + \alpha h \partial_z h + \beta \partial_z^3 h + \gamma \partial_z^5 h) = -(C_g/2)\partial_y^2 h\,, \qquad (2.52)$$

where h is the perturbation in the depth of the fluid, $\alpha = 1.5\, C_g/H_0$, $\beta = (C_g/\sigma)(H_0^2 - 3\sigma/\rho g)$ is the dispersion parameter, σ is the coefficient of surface tension, ρ is the fluid density, and

$$\gamma = (C_g/\sigma)\left[H_0^2(2H_0^2/5 - \sigma/\rho g) - (1/12)(3\sigma/\rho g - H_0^2)^2\right]. \quad (2.52)$$

One can see from the expression for β that the dispersion changes sign when $H_0 = (2\sigma/\rho g)^{1/2} \simeq 0.5$ cm (for pure water). A similar change in the dispersion also occurs for magnetosonic waves (see §1.2). In that case, however, this effect is associated with a change in the angle of propagation. As the depth decreases, β becomes so small that the higher-order derivatives can no longer be neglected. In dimensionless form (2.52) reduces to

$$\partial_\zeta(\partial_\tau\psi + \psi\partial_\zeta\psi + \varepsilon\partial_\zeta^3\psi + \partial_\zeta^5\psi) = -\partial_\eta^2\psi, \quad (2.53)$$

where ε is a dimensionless dispersion parameter proportional to β.

We seek a steady-state solution to (2.53) that travels at velocity $-u$. We represent ψ by the Fourier integral:

$$\psi = \int \psi_{\mathbf{k}} \exp(i\mathbf{kr} + ik_\zeta ut)d\mathbf{k}. \quad (2.54)$$

Fourier transformation of (2.53) then yields an integral equation $\psi_{\mathbf{k}}$

$$\psi_{\mathbf{k}} = G_{\mathbf{k}}N_{\mathbf{k}}, \quad N_{\mathbf{k}} = \frac{1}{2}\int \psi_{\mathbf{k}'}\psi_{\mathbf{k}-\mathbf{k}'}d\mathbf{k}', \quad (2.55)$$

where

$$G_{\mathbf{k}} = \frac{k_\zeta^2}{k_\zeta^2 u + k_\eta^2 - \varepsilon k_\zeta^4 + k_\zeta^6}.$$

(Cf. (2.50)). One can see from (2.55) that when $u > \varepsilon/4$, $G_{\mathbf{k}}$ is finite everywhere. For this reason equation (2.53) has solitary 2D solutions for these values of u.

Soliton solutions of (2.52) can only be found numerically, as has been done using the stabilizing-multiplier method.

Single-soliton and bisoliton solutions with the contour maps shown in Figures 2.5, 2.6 have been obtained. Figure 2.5 is an analogue of the single-soliton solution of the Kadomtsev–Petviashvili equation (see Figure 2.4), and Figure 2.6 is an analogue of the bisoliton solution.

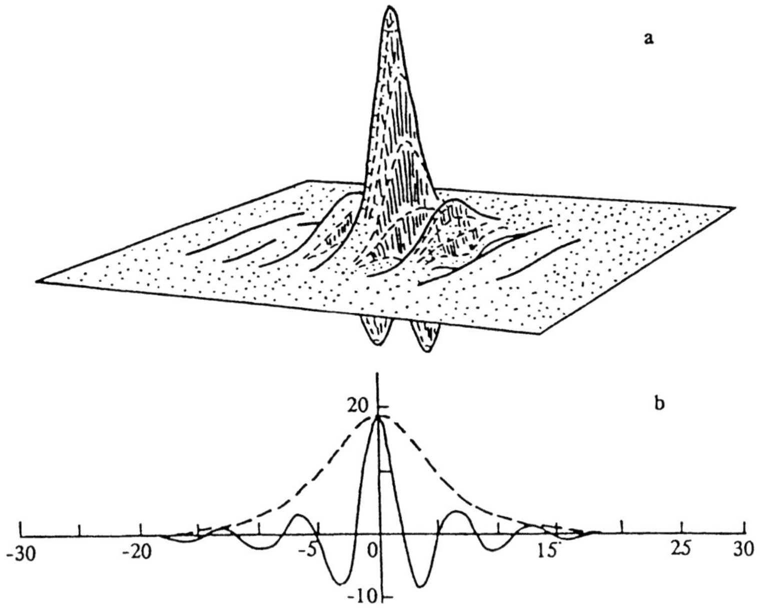

Figure 2.5. The 2D soliton solution of (2.53) for $\varepsilon = 1.9$: (a) general view, (b) principal cross-sections along (solid line) and perpendicular to (dashed line) the direction of motion.

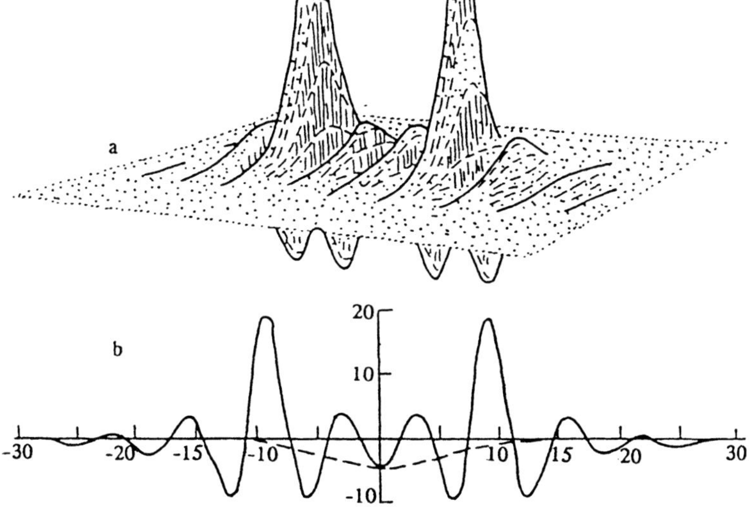

Figure 2.6. The 2D bisoliton solution of (2.53) at $\varepsilon = 1.9$: (a) general view, (b) principal cross-sections along (solid line) and perpendicular to (dashed line) the direction of motion.

2.6. MODEL EQUATION FOR ACOUSTIC WAVES IN AN ANISOTROPIC MEDIUM. 3D ION ACOUSTIC SOLITONS IN A MAGNETIC FIELD

In anisotropic media there is another preferred direction besides that in which a wave packet propagates. In the case of a plasma, this is the direction of the external magnetic field. Dispersion and diffraction effects become indistinguishable in such media. Two dispersion mechanisms appear. One is associated with the effect of Debye shielding and acts only along the magnetic field, the other is associated with the finiteness of the Larmor radius of the particles. This is well illustrated by the simple example of low-frequency ion acoustic waves (whose frequencies are low compared to ω_{Bi}). They are described by equation (1.14) in the linear approximation. The nonlinearity can, as in the preceding case, be found by neglecting dispersion. Assuming that the packet is shaped like a pancake with its axis ($k_z \gg k_\perp$) along the field, we get $v_z \gg v_\perp$. Proceeding as in the case of no magnetic-field, we find that when the amplitudes are small, the nonlinear term can be found by disregarding higher-order derivatives, so we again obtain (2.11). Combining (1.14) and (2.11), and recalling that Φ depends on arguments of the form (2.8), we get a nonlinear equation for low-frequency ion acoustic oscillations travelling along the magnetic field:

$$\partial_t \Phi + c_s \Phi \partial_z \Phi + (c_s r_D^2/2)\partial_z^3 \Phi = -(c_s r_s^2/2)\nabla_\perp^2 \partial_z \Phi . \quad (2.56)$$

This equation was derived and investigated by Zakharov and Kuznetsov (1974). Its dimensionless form is given in Appendix 1. Equation (2.56) has solutions in the form of a 1D soliton, (2.14). This solution can be examined for stability by the method explained in §2.2. One could have inferred from the sign of the right-hand side of (2.56) that the 1D soliton is unstable. We seek a 3D steady-state solution of (2.56):

$$\Phi = \Phi_0(\zeta, x, y), \quad \zeta \equiv z - ut. \quad (2.57)$$

Substituting (2.57) into the original equation (2.56) and integrating the result once yields a soliton equation with a quadratic nonlinearity. Its dimensionless form is

$$\nabla^2 \Phi = u\Phi - \Phi^2. \quad (2.58)$$

The equation has a spherically symmetric solution that is a function of

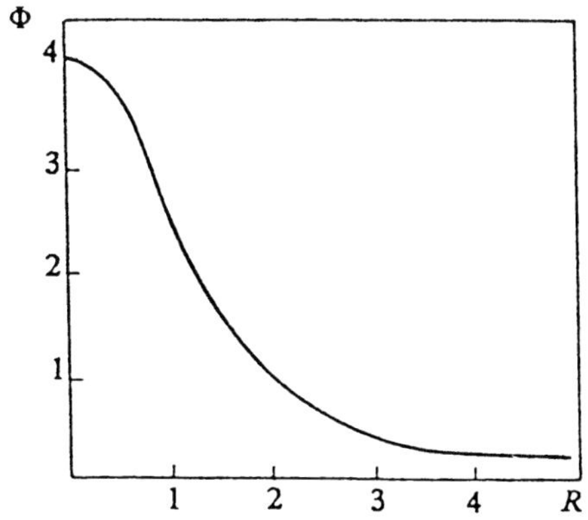

Figure 2.7. The spherically symmetric solution of (2.58) at $u = 1$.

of interest as a curious case of a 3D soliton which is stable. The stability was proved by Zakharov and Kuznetsov (1974). We shall present a proof based on the Lyapunov method. First, we point out that the dimensionless analogue (2.56) conserves the following integrals

$$P = \int \Phi^2 d^3x, \tag{2.59}$$

$$H = \int \left[(\nabla\Phi)^2/2 - (2/3)\Phi^3 \right] d^3x. \tag{2.60}$$

We construct the Lyapunov functional based on these functionals:

$$L = H + \lambda P. \tag{2.61}$$

The Lyapunov theory of stability is discussed in more detail in Appendix 2. For the moment, we just note that a sufficient condition for stability is that the solution realizes a minimum of H for fixed P. This reduces to the condition that the first variation of the functional should vanish, i.e. $\delta L = 0$. The first condition reduces to equation (2.58) and $\lambda = u$. In order to verify the second condition, we shall prove that L is bounded from below (Zakharov and Kuznetsov, 1974). Using Hölder's inequality

$$\left(\int \Phi^3 d^3x\right)^2 \leq \left(\int \Phi^2 d^3x\right) \left(\int \Phi^4 d^3x\right) \tag{2.62}$$

and Ladyzhenskaya's inequality

$$\left(\int \Phi^4 d^3x\right)^2 \leq \frac{4}{3}\left(\int \Phi^2 d^3x\right)\left(\int |\nabla\Phi|^2\, d^3x\right) \tag{2.63}$$

we get

$$L \geq -P^3/12 + uP. \tag{2.64}$$

As Euler's equation (2.58) has a unique solution, one concludes that L has a minimum just at the soliton solution where $\delta^2 L > 0$. Therefore, the soliton is stable.

2.7. NONLINEAR PROPAGATION OF ALFVÉN AND MAGNETOSONIC WAVES ALONG A MAGNETIC FIELD

As magnetohydrodynamic waves propagate along a magnetic field, their plane of polarization undergoes rotation: in the sense of the ion cyclotron rotation in the case of Alfvén waves and in the sense of the electron cyclotron rotation in the case of magnetosonic waves. Let us examine these waves when $\omega < \omega_{Bi}$, when the rotation velocity is small compared to ω. They can then be described by a simple model equation which we shall derive in this section. In the linear approximation a packet of these waves is described by the dispersion equation (1.31) in coordinate space. In Fourier space the components of the magnetic field are related by

$$(\omega^2 - k_z^2 C_A^2)B_x = (i\,\omega k_z^2 C_A^2/\omega_{Bi})B_y; \tag{2.65}$$

and

$$(\omega^2 - k_z^2 C_A^2)B_y = -(i\omega k_z^2 C_A^2/\omega_{Bi})B_x. \tag{2.66}$$

We have $B_x = -i\,\mathrm{sign}(\omega/\omega_{Bi})B_y$ in a Alfvén wave, and $B_x = +i\,\mathrm{sign}(\omega/\omega_{Bi})B_y$ in a magnetosonic wave. We shall only consider waves that are travelling in the positive z direction. Then from (2.65) and (2.66) we have approximately

$$(\omega - k_z C_A)B_x = (i/2)k_z^2 r_A^2 \omega_{Bi} B_y, \tag{2.67}$$

$$(\omega - k_z C_A)B_y = -(i/2)k_z^2 r_A^2 \omega_{Bi} B_x. \tag{2.68}$$

Choosing coordinates in a reference frame travelling at the Alfvén speed, we find that the equation corresponding to (2.67) and (2.68) is

$$(2/\omega_{Bi})\partial_t b^\sigma + i\sigma r_A \partial_z^2 b^\sigma = 0, \tag{2.69}$$

where $\sqrt{2}b^\sigma \equiv (B_x + i\sigma B_y)/B_0$, $\sigma = \pm 1$.

Note that, according to (2.67) and (2.68), each of the equations corresponding to $\sigma = \pm 1$ describes a superposition of clockwise and counterclockwise rotating waves.

Equation (2.69) should be supplemented by a nonlinear term in accordance with the Korteweg–de Vries scheme. We can disregard dispersion effects when calculating the nonlinear term and use the equations of one-fluid MHD:

$$(\partial_t + v_z\partial_z)\mathbf{v}_\perp = C_A^2\partial_z(\mathbf{B}_\perp/B_0), \tag{2.70}$$

$$(\partial_t + v_z\partial_z)v_z = -(C_A^2/2)\partial_z(\mathbf{B}_\perp/B_0)^2, \tag{2.71}$$

$$\partial_t\mathbf{B}_\perp = \partial_z(\mathbf{v}_\perp B_0 - v_z\mathbf{B}_\perp), \tag{2.72}$$

$$\partial_t\rho + \partial_z(v_z\rho) = 0. \tag{2.73}$$

The dependence of all the quantities on time and z is chosen to be in the form $\rho = \rho(z - C_At, t)$ and so on. The dependence on the second argument is due to nonlinear effects. Thus, it can be assumed to be much weaker than the dependence on the first argument. The perturbations in ρ and v_z are due to the nonlinearity and are, therefore, small. The dependence on the second argument can be neglected in them. One can then set $\partial_t \simeq C_A\partial_z$ in (2.73). Integrating with respect to z, we get

$$\rho = \rho_0(1 + v_z/C_A). \tag{2.74}$$

Here we have used the fact that $\rho = \rho_0$ at infinity. The term v_z^2 on the left side of (2.71) can be neglected, to yield

$$v_z = (C_A/2)(\mathbf{B}_\perp/B_0)^2. \tag{2.75}$$

Further, on substituting (2.74) and (2.75) into (2.70) and (2.72) we obtain

$$\left[\partial_t - C_A\partial_z - (C_A/2)\partial_z(\mathbf{B}_\perp/B_0)^2\right]\mathbf{B}_\perp = B_0\partial_z\mathbf{v}_\perp, \tag{2.76}$$

$$B_0\left[\partial_t - C_A\partial_z + (C_A/2)(\mathbf{B}_\perp/B_0)^2\partial_z\right]\mathbf{v}_\perp$$
$$= C_A^2\partial_z\mathbf{B}_\perp - (C_A^2/2)(\mathbf{B}_\perp/B_0)^2\partial_z\mathbf{B}_\perp. \tag{2.77}$$

Equating terms of the same order in the linear approximation in $B_\perp$ in (2.74) and (2.77), we get

$$\mathbf{v}_\perp = -C_A(\mathbf{B}_\perp/B_0). \tag{2.78}$$

The next order in $\mathbf{B}_\perp$ yields a nonlinear equation for the temporal evolution of $\mathbf{B}_\perp$ from (2.76) when the effects of dispersion and diffraction are neglected:

$$\partial_t\mathbf{B}_\perp + (C_A/2)\partial_z|\mathbf{B}_\perp|^2\mathbf{B}_\perp = 0. \tag{2.79}$$

Combining the parabolic equation (2.69) with (2.79), which involves a nonlinear term, we obtain a set of dimensionless equations for the longitudinal propagation of Alfvén and magnetosonic waves (Rogister, 1971, Mio *et al.*, 1976):

$$\partial_t b^\sigma + \partial_z|b|^2 b^\sigma + i\partial_z^2 b^\sigma = 0, \tag{2.80}$$

where $|b|^2 = b^+b^- = b_x^2 + b_y^2$.

From this one can see that the Alfvén and magnetosonic waves interact because of the nonlinearity. The physical cause of this phenomenon lies in the fact that the propagation velocity depends on the high frequency pressure $(\mathbf{B}_\perp)^2$ which involves contributions from both types of waves.

Under certain conditions these may become coupled and form a steady-state soliton, in spite of the fact that their group velocities are different in the linear approximation. A soliton of this kind was found by Mio *et al.* (1976):

$$b^\sigma = (A/2)^{1/2}\left[\exp(-A\zeta) + i\sigma\exp(A\zeta)\right]$$
$$\times\exp(-i\sigma A^2\tau)\,\mathrm{ch}^{-2}(2A\zeta), \tag{2.81}$$

where A is the soliton amplitude.

Kaup and Newell (1978) have shown that (2.81) can be fully integrated and solved by the ISP method. The 1D soliton of this type has not yet been studied for stability with respect to perturbations in the front. Note that, if it were purely magnetosonic, it would be unstable, because its dispersion is positive.

2.8. PROPAGATION OF MAGNETOSONIC WAVES AT AN ANGLE TO A MAGNETIC FIELD

When the Alfvén and magnetosonic waves propagate obliquely to the magnetic field, their phase velocities become appreciably different. The Alfvén wave then becomes anisotropic, while the anisotropic corrections to the magnetic sound are small at frequencies below ω_{Bi} and show up only in the dispersion term. When the propagation occurs nearly transverse to the magnetic field, these corrections are also small when $\omega > \omega_{Bi}$ (see, for example, (1.29)). It follows from the preceding section that the nonlinearity in the magnetic sound is cubic in the amplitude for strictly longitudinal propagation. However, propagation at a finite angle gives rise to a quadratic nonlinearity of the advective type (Korteweg–de Vries type) associated with the $(\mathbf{v}\cdot\nabla)\mathbf{v}$ term in the equation of motion and with other similar terms. We shall derive a simplified form of this nonlinearity by disregarding the corrections for dispersion and diffraction. The starting point is the equations of one-fluid MHD:

$$d_t\mathbf{v} = \frac{1}{4\pi\rho}(\text{curl}\ \mathbf{B}\times\mathbf{B}), \tag{2.82}$$

$$\partial_t\mathbf{B} = \text{curl}\ (\mathbf{v}\times\mathbf{B}), \tag{2.83}$$

and

$$\partial_t\rho - \text{div}\,\rho\mathbf{v} = 0. \tag{2.84}$$

We have neglected pressure oscillations compared with those in the magnetic field, which is correct for small β. The plasma conductivity is assumed to be infinite; hence, the electric field is given by $\mathbf{E} = -(\mathbf{v}\times\mathbf{B})/c$ in this approximation. Suppose there is a 1D wave travelling at an angle of $\theta = \theta_0$ to the ambient magnetic field which is in the z direction. We seek the perturbed quantities as functions of the coordinates and time in the form:

$$\rho = \rho(\zeta, \tau),\quad \zeta \equiv z\cos\theta_0 + x\sin\theta_0 - C_At,$$
$$\tau = t. \tag{2.85}$$

We thus assume that the packet depends much more strongly on ζ than on the coordinates that are perpendicular to ζ. Proceeding by analogy with the derivation of the Kadomtsev–Petviashvili equation, we obtain the simplified nonlinear equation

$$\partial_\tau b + (3/2)C_A b\partial_\zeta b = 0, \quad b \equiv (B_z - B_0)/B_0. \tag{2.86}$$

Here b is the dimensionless perturbation component of the magnetic field along the z axis.

The other quadratic nonlinear terms can be disregarded, because they involve small derivatives with respect to the transverse coordinates. We shall only include them in the linear approximation.

We now add dispersion and diffraction terms to (2.86). To do this, we define coordinates perpendicular to ζ:

$$\xi = z \sin\theta_0 - x\cos\theta_0, \quad \eta = y.$$

We shall work in the Leontovich approximation $k \simeq k_\zeta + (k_\xi^2 + k_\eta^2)/2k_\zeta$. Then, using the linear dispersion relation (1.32), we can reduce (2.86) to the form

$$\partial_\zeta(C_A^{-1}\partial_\tau b + 1.5b\partial_\zeta b + r_A^2\alpha_0\partial_\zeta^3 b) = -0.5\nabla_\perp^2 b,$$
$$\alpha_0 = 0.5\cot^2\theta_0 \gg \beta_i + m_e/m_i. \tag{2.87}$$

Thus, a magnetosonic wave packet propagating at angles that are not near either 0 or $\pi/2$ obeys the Kadomtsev–Petviashvili equation with positive dispersion. Recalling the results of the preceding discussion, one may conclude that oblique propagating magnetosonic waves do not form stable solitons. Packets of such waves either collapse or spread out. In the presence of a steady-state source, one can expect self-focusing in a beam of such waves. It has been pointed out (Sagdeev *et al.*, 1977) that collapse can act as an effective mechanism for dissipation in oblique shock waves of the magnetosonic type. Since collapse occurs only at frequencies below ω_{Bi}, while a shock wave travels at a velocity close to C_A, it follows that the width of the shock front must be large compared to r_A.

The sign of the dispersion changes for angles of propagation close to $\pi/2$. In that case, dispersion remains weak at frequencies above ω_{Bi}.

At these frequencies, extra diffraction in the direction of the constant field becomes significant. This effect is accommodated by expanding (1.31) around $\theta = \theta_0$. We then obtain

$$\omega = C_A k_\zeta \left[1 + k_\zeta^2 r_A^2(\alpha_* + \alpha_1\theta_1 + \alpha_2\theta^2) + C_A k_\perp^2/2k_\zeta \right], \quad (2.88)$$

where $\theta_1 = \theta - \theta_0 \simeq k_\xi/k_\zeta$, $\alpha_1 = \partial_{\theta_0}\alpha_0$, $\alpha_2 = 0.5\partial_{\theta_0}^2\alpha_0$, and $\alpha_* = \alpha_0 - \beta_i/4 - m_e/2m_i$. Combining (2.88) and (2.86), we find that the sign of the dispersion in (2.88) can change and two new terms involving mixed derivatives are to be added:

$$\partial_\zeta(\partial_\tau b + 1.5b\partial_\zeta b + r_A^2\alpha_*\partial_\zeta^3 b) = -0.5\nabla_\perp^2 b + \alpha_1\partial_\xi\partial_\zeta^3 b + \alpha_2\partial_\zeta^2\partial_\xi^2 b. \quad (2.89)$$

Depending on the sign of α_*, this equation can have 1D, 2D, and 3D soliton solutions. When $\alpha_* < 0$, only 1D solitons exist. Sagdeev (1966) was the first to derive these. He also noted that for propagation within the cone of angles $\tan^2\theta_0 < m_e/m_i + \beta_i/4$, where the magnetic sound has negative dispersion, the soliton is a wave of density and magnetic field compression, while outside the cone it is a rarefaction wave, that is, the magnetic field and density are smaller within a soliton than in the surrounding background. Sagdeev's solutions coincide with the 1D solutions of the Kadomtsev–Petviashvili equation for small amplitudes. Later it became clear that 1D solitons with rarefaction are unstable with respect to curvature of the soliton front, like solitons in media with positive dispersion. 2D solitons can exist for $\alpha_* > 0$. When mixed-derivative corrections to the Kadomtsev–Petviashvili equation are insignificant, 2D solitons are unstable, as has been shown above. It is shown by Mikhailovskii *et al.* (1985) that the corrections can stabilize 2D solitons. We shall establish when this is possible. It turns out that this occurs for frequencies $\omega \gtrsim \omega_{Bi}$ over sizes smaller than r_A, and at angles of propagation sufficiently close to that at which the dispersion changes sign.

2D soliton solutions of (2.89) have the same form as the rational soliton solutions of the Kadomtsev–Petviashvili equation. Their stability will be investigated as in §2.2. Linearizing (2.89) around the soliton solution (2.27), we get

$$L\psi = i\partial_\zeta(\omega\psi - \alpha_1 k\partial_\zeta^3\psi) - k^2(\psi - \alpha_2\partial_\zeta^2\psi), \quad (2.90)$$

where the operator L is defined by (2.33).

Expanding ψ in a series of powers of the small parameter $\omega \sim k$ in a similar fashion to (2.16), we obtain a sequence of equations similar to (2.17)–(2.19) with some extra terms:

$$L\psi_0 = 0; \tag{2.91}$$

$$L\psi_1 = i\omega\partial_\zeta\psi_0 - i\alpha_1 k\partial_\zeta^3\psi_0; \tag{2.92}$$

$$L\psi_2 = i\omega\partial_\zeta\psi_1 - i\alpha_1 k\partial_\zeta^3\psi_1 - k^2(\psi_0 - \alpha_2\partial_\zeta^2\psi_0). \tag{2.93}$$

The zeroth-order solution (2.91) coincides with (2.20). Substituting it into (2.92), we get

$$\psi_1 = i\omega\partial_u\psi_0 + \frac{ik\alpha_1}{2\sigma}\partial_{\alpha_0}\psi_0. \tag{2.94}$$

Using this, we obtain an equation for ψ_2. Proceeding in a similar fashion to § 2.2, we find the condition under which this equation can be solved

$$\omega^2 + \alpha_1\alpha_*^{-1}u\,\omega(1 - 3I) + 4u + 4u^2\alpha_*^{-1}I = 0, \tag{2.95}$$

where

$$I = \frac{\int(\partial_\zeta\Phi_0)^2 d\zeta dx}{\int \Phi_0^2 d\zeta dx}. \tag{2.96}$$

One can show that $I = 2$ using the soliton equation (2.27). From (2.95) one then derives the conditions for stability of the 2D soliton, i.e., the roots of (2.95) must be real,

$$(\alpha_1/\alpha_*)^2 u > 16/25 + (32/25)u\alpha_*^{-1}. \tag{2.97}$$

This determines the range of angles for which a 2D soliton is stable

$$\beta_i/4 + m_e/2m_i < \cos^2\theta_0 < 3/10. \tag{2.98}$$

The velocity of propagation u must also be large enough, in accordance with (2.97).

To sum up, magnetosonic waves exhibit different nonlinear properties for different angles of propagation. This is related to the sign of the dispersion and to the type of nonlinearity.

3. Nonlinear equations of the optical type

3.1. NONLINEAR SCHRÖDINGER EQUATION

The nonlinear Schrödinger equation seems to be one of the most important nonlinear equations, to judge from its applications in plasma wave theory and in other fields of physics. Equations like Schrödinger's equation are of particular interest in the theory of elementary particles.

Schrödinger's equation can be derived, for example, from the Klein–Gordon nonlinear equation under the assumption that the ratio of kinetic energy to rest energy and the particle number density are both small. The latter equation has the form

$$\partial_t^2\varphi - \nabla^2\varphi + (1 + M)\,\varphi = 0. \tag{3.1}$$

Here $M = M(|\varphi|)$ is a nonlinear real-valued function which incorporates corrections to the rest mass (taken to be unity here). For simplicity all external fields are assumed to be absent. The principal qualitative difference between (3.1) and equations of the acoustic type consists in the presence of a third term incorporating the effect of the rest mass. For large wavelengths, therefore, the wave packet frequencies no longer vanish, but approach $\pm\sqrt{1 + M}$. Even though this equation has real coefficients, it can have a complex solution. Two integrals are conserved, the energy H and charge Q,

$$H = \frac{1}{2}\int \left\{|\partial_t\varphi|^2 + |\nabla\varphi|^2 + |\varphi|^2\right\}d^3x + H_M, \tag{3.2}$$

$$Q = i\int (\varphi^*\partial_t\varphi - \varphi\partial_t\varphi^*)d^3x. \tag{3.3}$$

where H_M is the part of energy related to M.

Equation (3.1) has an oscillating soliton solution of the form $\varphi = f(r) \exp(-i\omega t)$, where f vanishes at infinity and satisfies the equation

$$\nabla^2 f = (1 - \omega^2)f - Mf. \tag{3.4}$$

The soliton solutions of (3.4) are spherically symmetric and exist only if $\omega^2 < 1$ and $M > 0$. The stability of these solitons has been studied by Zastavenko (1965). It turns out that if $M = |\varphi|^{2n}$, then m-dimensional soliton solutions in m-dimensional space are stable, provided

$$mn/2 < \omega^2 < 1, \quad n > 0. \tag{3.5}$$

From this one can see that when the amplitude increases (which is easily verified to correspond to a decrease in ω^2), the soliton is no longer stable.

In the limit of small amplitude and small $|\nabla^2 \varphi|$ (the corresponding assumption in relativity theory is that the kinetic energy is small compared to the rest energy), equation (3.1) yields Schrödinger's equation. It is derived by representing the solution of (3.1) as an oscillating function of unit frequency whose amplitude is weakly dependent on the time and coordinates:

$$\varphi = \frac{1}{2}\left[\psi(r, t) \exp(-it) + c.c.\right]. \tag{3.6}$$

Expanding in powers of ψ and the kinetic energy in the parabolic approximation ($\nabla^2\psi/\psi \ll 1$), we get Schrödinger's equation, apart from some numerical coefficients.

$$i\partial_t \psi + \nabla^2 \psi - V(p)\psi = 0, \quad p = \psi\,\psi^*. \tag{3.7}$$

Here V is a nonlinear potential for Schrödinger's equation corresponding to the nonlinear correction to the rest mass in (3.1).

This equation has the following integrals: the number of particles

$$N = \int |\psi|^2 d^3x \tag{3.8}$$

and the energy

$$H = \int \left[|\nabla\psi|^2 + \int V dp \right] d^3x. \tag{3.9}$$

The conservation of energy follows from the Hamiltonian form of (3.7)

$$i\partial_t\psi = \delta H/\delta\psi^*. \tag{3.10}$$

In the general case the steady-state solutions of (3.7) are of the form

$$\psi = \psi_0(\mathbf{r} - 2\mathbf{k}t)\sigma,$$
$$\sigma = \exp\left\{ i\left[-(\omega_0 - k^2)t + \mathbf{kr}\right]\right\}, \tag{3.11}$$

where $\mathbf{k}$ is an arbitrary constant vector and ω_0 is the frequency. Substituting (3.11) into (3.7) yields an equation which must be satisfied by ψ_0:

$$\nabla^2\psi_0 = -\left[\omega_0 - V(p_0)\right]\psi_0, \quad p_0 = \psi_0^2; \tag{3.12}$$

hence, one can see that ψ_0 is a real-valued function. A dependence of ψ_0 on the coordinates is called amplitude modulation, while the factor σ provides phase modulation. The equation may have solutions in the form of 3D periodic waves and spherically symmetric 3D solitons. We shall examine the soliton solutions of (3.12). For a soliton solution to exist, there must be a potential well in the region of small p_0; this occurs if $V' < 0$ in this vicinity. The prime denotes the derivative with respect to p_0. We set $V(0) = 0$ for convenience. The soliton energy is negative in a well of this kind. We introduce the notation $\omega_0 = -A^2$. A soliton solution is stable if the Hamiltonian H has a minimum for this solution with the constraint that the number of particles should be constant. Small departures from a soliton solution would then not grow indefinitely, because the relevant departure of the Hamiltonian from the minimum as a functional of $\delta\psi$ is positive. We now find those values of V for which the Hamiltonian can have a relative minimum for the steady-state solutions of (3.12). Obviously, a sufficient condition for this to occur is that the second variation of H should be positive for all departures of ψ from the steady-state solution that conserve N. We construct the Lyapunov functional

$$L = A^2N + H. \tag{3.13}$$

Equating the first variation of the functional (3.13) to zero yields (3.12), where $\omega_0 = -A^2$. We write the variation in the form

$$\psi = \psi_0 + \delta\psi, \quad \delta\psi = \varphi_1 + i\,\varphi_2, \tag{3.14}$$

where φ_1, φ_2 are real-valued functions, as is ψ_0. The phase factor in (3.11) does not affect the result, so it can be omitted. Substituting (3.14) into (3.13) and using (3.12), we get

$$\delta^2 L = \int (\varphi_1 L_1 \varphi_1 + \varphi_2 L_0 \varphi_2) d^3x, \tag{3.15}$$

where L_0, L_1 are the operators

$$L_0 = -\nabla^2 + V_0 + A^2; \quad V_0 = V(p_0),$$
$$L_1 = L_0 + V_0' p_0. \tag{3.16}$$

To a first approximation the constraint of constant N in the variation of (3.14) yields

$$\int \varphi_1 \psi_0 d^3x = 0. \tag{3.17}$$

Comparing (3.12) and (3.16) yields

$$L_0 \psi_0 = 0; \quad L_1 \nabla \psi_0 = 0,$$
$$L_1 \partial_A \psi_0 = -2A\psi_0. \tag{3.18}$$

One can see that ψ_0 is a nodeless eigenfunction of L_0, so that it is a ground state function with the minimum energy, an eigenvalue of the operator. For an arbitrary function, therefore, we have

$$\int \varphi_2 L_0 \varphi_2 d^3x \geq 0. \tag{3.19}$$

It is now sufficient to demonstrate that the integral to the first term on the right-hand side of (3.15) is positive, and then the solution can be considered stable. From (3.16) one can see that L_1 differs from L_0 in having a negative term added to it; hence, if it were not for the condition (3.17), the expression (3.15) would certainly have been negative, making the soliton unstable.

L_1 is a Hermitian operator; therefore, it has a complete set of eigenfunctions χ_ω with eigenvalues ω that satisfy

$$L_1 \chi_\omega = \omega\chi_\omega, \quad \int \chi_\omega \chi_{\omega'} d^3x = \delta(\omega - \omega'). \tag{3.20}$$

Expanding ψ_0 and φ_1 in terms of these orthogonal functions as

$$\psi_0 = \sum_\omega \psi_\omega \chi_\omega, \quad \varphi_1 = \sum_\omega \varphi_\omega \chi_\omega \,. \tag{3.21}$$

and substituting these expansions into (3.15), we get

$$\delta^2 L \geq \sum_\omega \omega |\varphi_\omega|^2, \tag{3.22}$$

where (3.19) and (3.20) have been used.

Using (3.20), (3.21), we deduce from (3.17) that the spectrum of φ is not arbitrary, but must satisfy the condition of constant N:

$$\sum_\omega \varphi_\omega \psi_\omega = 0. \tag{3.23}$$

According to the first of equations (3.18), L_1 has an eigenfunction $\nabla\psi_0$ with a zero eigenvalue $\omega = 0$. Since we are considering a nodeless solution ψ_0 to (3.18), it follows that when $\omega = 0$, the operator L_1 has solutions with a single node. Hence, it follows that L_1 has a nodeless solution χ_g with an eigenvalue $\omega = \omega_g < 0$, while all the other eigenvalues are positive. Substituting φ_g from (3.23) into (3.22), we get

$$\delta^2 L \geq \sum_{\omega>0} \omega |\varphi_\omega|^2 + \left(\omega_g / |\psi_g|^2\right) \left(\sum_{\omega>0} \varphi_\omega \psi_\omega\right)^2 . \tag{3.24}$$

Here ψ_g is the coefficient in front of the ground state function χ_g in the expansion of ψ_0. Note that because ψ_0 is orthogonal to the eigenfunction corresponding to $\omega = 0$, which according to (3.18) is $\nabla\psi_0$, we have $\psi_\omega \big|_{\omega=0} = 0$.

Hölder's inequality yields the estimate

$$\left(\sum_{\omega>0} \varphi_\omega \psi_\omega\right)^2 \leq \left(\sum_{\omega>0} \omega |\varphi_\omega|\right)^2 \left(\sum_{\omega>0} |\varphi_\omega|^2/\omega\right). \tag{3.25}$$

Substituting (3.25) into (3.24), we get a simpler inequality:

$$\delta^2 L \geq -a^2 \left[|\psi_g|^2/\omega_g + \sum_{\omega>0} |\psi_\omega|^2/\omega \right]. \tag{3.26}$$

Here we have $a^2 > 0$, since $\omega_g < 0$. We now expand the last of equations (3.18) in the eigenfunctions (3.20):

$$\omega \partial_A \psi_\omega = -2A\, \psi_\omega. \tag{3.27}$$

Taking this equality into account, we transform (3.26) to

$$\delta^2 L \geq \frac{a^2}{2A} \sum_\omega \psi_\omega^* \partial_A \psi_\omega = \frac{c^2}{4A} \partial_A \sum_\omega |\psi_\omega|^2, \tag{3.28}$$

the sum being taken over all the eigenstates. Using the first expansion in (3.21) and the expression for the number of particles (3.8), from (3.28) we obtain the condition for stability of the soliton solution (the second variation of L must be positive):

$$\partial_A N > 0. \tag{3.29}$$

This criterion is due to Vakhitov and Kolokolov (1973). When expressed in simple words, its meaning is as follows: a sufficient condition for stability is that the number of particles in a soliton, N, should increase as its amplitude A increases.

Consider the particular case $V = -p_0^n$. Then, remembering that $\omega_0 = -A^2$, we get $N \sim A^{2/n-m}$, where m is the dimensionality of the space. From (3.29) it follows that a soliton is stable for a power-law potential in an m-dimensional space, provided $mn < 2$ (cf. (3.5)). Otherwise, a soliton solution either collapses or spreads, depending on the form of the initial perturbation. Vakhitov and Kolokolov (1973) have examined the soliton solutions of (3.12) for $V = -p_0/(1 + p_0)$. This is a nonlinear potential in which the nonlinearity saturates with increasing amplitude. The solution is unstable for small amplitudes, because saturation is not attained and, in effect, $n = 1$. As A increases, the degree of nonlinearity n actually decreases, and the soliton should become stable. Figure 3.1 shows a plot of N as a function of A^2. It can be seen that a 3D soliton becomes stable for $A^2 > 0.08$ with this kind of potential. The criterion (3.29) seems to be more general than implied by the case we

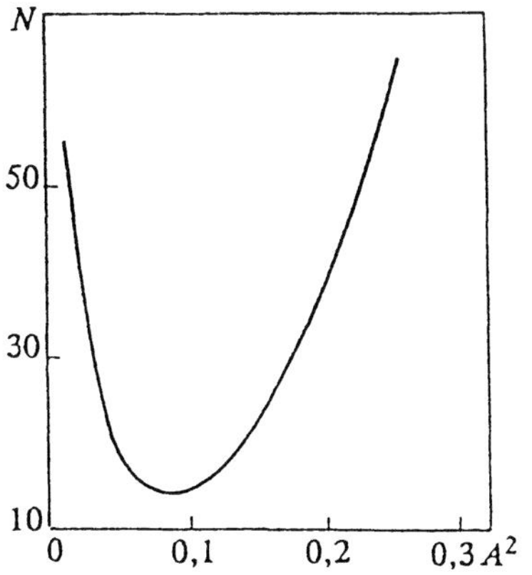

Figure 3.1. Particle number N as a function of amplitude A for a nodeless soliton solution of (3.12) with saturated nonlinearity.

have proved here. This is true, for example, for complex operators and departures from spherical symmetry. In such cases it sometimes happens that integrals of motion exist, but there is no rigorous proof of the existence of stable-state solutions.

3.2. LANGMUIR WAVES IN AN ISOTROPIC PLASMA. ZAKHAROV'S EQUATION

In the steady state a charge in a plasma is shielded over distance on the order of Debye length $r_D = v_{Te}/\omega_{pe}$. However, high-frequency oscillations of the charge (at frequencies $\gtrsim\omega_{pe}$) are no longer shielded, and propagate as the waves that were discovered by Langmuir in 1930. The dispersion relation for Langmuir waves has the form

$$\omega^2 = \omega_{pe}^2(1 + 1.5k^2r_D^2), \quad kr_D \ll 1. \tag{3.30}$$

These waves are readily excited by beams of electrons or electromagnetic waves because of their small phase velocity. For this reason they play an important role in beam and *RF* heating of plasmas. The energy density of Langmuir waves W exceeds that of the plasma, nT, in a number of experiments. This is a significant effect in, for example, determining the temperature of laser plasmas and Z-pinches from spectral line broadening. However, highly nonlinear effects appear much earlier with

Langmuir waves, when $W/nT \gtrsim k^2 r_D^2$ (Vedenov and Rudakov, 1965). At this level of turbulence, the approximation of chaotic phases for the Fourier components of the electric field is no longer valid. Zakharov (1972) proposed a simplified equation for nonlinear Langmuir waves. Its main nonlinearity is due to the high frequency pressure, which affects the ion density. This produces density minima (cavities) in which the high frequency field is trapped. Zakharov (1972) has shown that these cavities are unstable and trap the high frequency field by collapsing to a size on the order of the Debye length, where electron Landau damping becomes important. This phenomenon is referred to as Langmuir collapse.

When a source of energy is available, the Landau damping in cavities can be compensated, to lead to the formation of steady-state cavity-Langmuir solitons (Petviashvili and Tsvelodub, 1980). Thus, strong Langmuir turbulence may involve the spontaneous production and collapse of cavities or of quasi-steady-state dissipative solitons, depending on the kind of source.

We now proceed to derive Zakharov's equation (Zakharov, 1972). For simplicity we assume $T_i \simeq T_e$, so that the ion acoustic waves are subject to strong attenuation. We separate the oscillating field into high-frequency and low-frequency parts:

$$\varphi = \frac{1}{2}\left\{\psi\,(r,\,t)\exp\left(-i\,\omega_p t\right) + c.c.\right\} + \Phi(r,\,t). \qquad (3.31)$$

Here ψ, the amplitude of the high-frequency oscillations, is a slowly varying function of time ($d_t\,\psi \ll \omega_{pe}\psi$) and Φ is the low-frequency part of the potential. The equations of motion for the electrons will be divided into slowly varying and rapidly oscillating parts. Averaging over the rapid oscillations, we get

$$\langle \mathrm{v}_e \cdot \nabla \mathrm{v}_e \rangle = (e/m_e)\nabla\Phi - (T_e/m_e)\nabla(\delta n/n_0), \qquad (3.32)$$

where δn is a low-frequency perturbation in the plasma density and the angular brackets denote averaging over the rapid oscillations. The dependence of ψ on time is assumed to be weak. Since the ion acoustic waves are strongly attenuated, the ions are distributed in accordance with a Boltzmann distribution in the low-frequency potential Φ

$$\delta n \simeq -(e\,\Phi/T_i)n_0. \qquad (3.33)$$

Departures from quasi-neutrality can be disregarded in low-frequency oscillations, so that the low-frequency part of the electron density is also given by (3.33).

Using (3.31) from the equation of electron motion we obtain

$$\mathbf{V}_e = \frac{ie}{2m_e \omega_{pe}} (1 + 1.5r_D^2\nabla^2)\nabla\psi \exp(-i\omega_{pe}t) + c.c. \tag{3.34}$$

We use Maxwell's equation

$$\partial_t \nabla^2 \varphi = 4\pi \operatorname{div} \mathbf{j},$$

where

$$\mathbf{j} = -e(n_0 + \delta n)\mathbf{V}_e, \tag{3.35}$$

to close this system of equations.

Separating the high-frequency part and substituting (3.31), (3.33), and (3.34) in (3.35), we get

$$\nabla^2\left(\frac{2i}{\omega_{pe}} \partial_t\psi + 1.5r_D^2\nabla^2\psi\right) + \frac{\operatorname{div}(|\nabla\psi|^2\nabla\psi)}{16\pi n_0(T_i + T_e)} = 0. \tag{3.36}$$

In dimensionless form this equation becomes

$$\nabla^2(i\partial_t\psi + \nabla^2\psi) + \operatorname{div}(|\nabla\psi|^2\nabla\psi) = 0. \tag{3.37}$$

This equation has been obtained by Rudakov (1973) for the 1D case. Like Schrödinger's equation, this equation also conserves the energy and the number of particles

$$H = \int (|\nabla^2\psi|^2 - 0.5|\nabla\psi|^4)d^3x, \tag{3.38}$$

$$N = \int |\nabla\psi|^2 d^3x. \tag{3.39}$$

These integrals can be used to study the stability of the soliton solutions in a fashion similar to that used for Schrödinger's equation. Equation (3.37) has steady-state solutions that oscillate at a low frequency, $\omega = -A^2$:

$$\psi = \frac{1}{2}\left\{\psi_0(r) \exp(i A^2 t) + c.c.\right\} \tag{3.40}$$

Substituting (3.40) into (3.37), we obtain an equation for the real-valued function $\psi_0(r)$:

$$\nabla^2(\nabla^2\psi_0 - A^2\psi_0) + \operatorname{div}(|\nabla\psi_0|^2\nabla\psi_0) = 0. \tag{3.41}$$

This equation has solutions in the form of 1D, cylindrical, and spherically symmetric solitons. Proceeding in a similar manner to the preceding discussion, one can show that all these solutions are unstable. To see this, we substitute (3.40) into (3.39):

$$N = \text{const}\, A^{2-m}. \tag{3.42}$$

Here m is the dimensionality of the space.

According to the Vakhitov–Kolokolov criterion (3.29), m-dimensional soliton solutions of (3.41) in a space with the same number of dimensions are unstable, except in the one-dimensional case, where $m = 1$. However, this solution too is unstable with respect to 2D and 3D perturbations. This can be demonstrated in a fashion similar to that used for studying the stability of the 1D soliton of the Kadomtsev–Petviashvili equation (see also § 3.3).

Multiplying (3.42) by $r \cdot \nabla\psi_0$, integrating over the volume, and using the localization property of ψ_0, we get

$$H = \left[(m-2)/(4-m)\right] A^2 N \tag{3.44}$$

Hence, one can see that the soliton energy is negative in 1D, vanishes in 2D, and is positive in 3D space.

3.3. EQUATION FOR LANGMUIR WAVES IN A MAGNETIC FIELD. STEADY-STATE SOLUTIONS AND THEIR STABILITY

In the previous section we derived the equation for Langmuir waves without a magnetic field, but including the most important nonlinear effect, the ponderomotive force. The typical dispersion length was equal to the Debye radius. The presence of a weak magnetic field greatly increases this length in a direction transverse to the magnetic field. We now consider the resulting effect on the equation for Langmuir waves.

The dispersion relation for electrostatic (potential) waves with a frequency close to ω_{pe} and a wave vector in a direction close to that of the magnetic field has the form

$$\omega^2 = \omega_{pe}^2(1 + 1.5k_z^2 r_D^2 + \alpha k_\perp^2 / k_z^2),$$
$$\alpha k_\perp^2 \ll k_z^2, \quad k^2 r_D^2 \ll 1, \tag{3.44}$$
$$\alpha \equiv \omega_{Be}^2 / (\omega_{pe}^2 - \omega_{Be}^2).$$

Here ω_{pe} and ω_{Be} are the electron Langmuir (plasma) and cyclotron frequencies. If the packet frequency is around ω_{pe}, then the potential in it can be represented in the form (3.31).

For simplicity we assume that the packet velocity is small compared to the ion acoustic speed. In that case, the change in the plasma density due to the high frequency force can be expressed as

$$\delta n = -\frac{1}{16\,\pi(T_i + T_e)} |\partial_z \psi|^2. \tag{3.45}$$

The electric field perpendicular to the magnetic field can be neglected in (3.45), because ψ is weakly dependent on $r_\perp$. From this we find a local change in the Langmuir frequency, and then equations (3.44) and (3.45) yield (Petviashvili, 1975):

$$\partial_z^2\left(\frac{2i}{\omega_{pe}} \partial_t \psi + 1.5 r_D^2 \partial_z^2 \psi\right) - \partial_z \left(\frac{\delta n}{n_0} \partial_z \psi\right) = \alpha \nabla_\perp^2 \psi,$$
$$\nabla_\perp^2 \equiv \partial_x^2 + \partial_y^2, \quad \nabla_\perp^2 \ll \partial_z^2 \psi, \tag{3.46}$$

where n_0 is the mean plasma density. Equation (3.46) may have different signs for the transverse dispersion, depending on the sign of α. When the packet has finite dimensions, (3.46) conserves the number of particles N and energy H in the form

$$N = \int |\partial_z \psi|^2 d^3x, \tag{3.47}$$

$$H = \int \left[1.5 r_D^2 |\partial_z^2 \psi|^2 + \alpha |\nabla_\perp \psi|^2 - \frac{|\partial_z \psi|^4}{32\,\pi n_0 (T_i + T_e)}\right] d^3x. \tag{3.48}$$

In accordance with the sign of α in (3.44), equation (3.46) can have a steady-state 3D solution, if the plasma frequency is greater than the cyclotron frequency, when the contribution to the energy (3.48) from the transverse electric field is positive. We seek a steady-state solution of (3.46) in the form

$$\partial_z \psi = (32\,\pi n_0(T_i + T_e))^{1/2} A\, f(\rho, \zeta) \exp\,(i\,\omega_{pe} A^2 t/2). \quad (3.49)$$

Here we introduce the dimensionless coordinates

$$\zeta = \frac{A}{3^{1/2}} \frac{z}{r_D}, \quad \rho = \frac{A^2}{(3\alpha)^{1/2}} \frac{r_\perp}{r_D}. \quad (3.50)$$

We then obtain the following equation for the dimensionless real-valued function f:

$$\partial_\zeta^2 \left(\partial_\zeta^2 f - f + f^3 \right) = \nabla_\rho^2 f. \quad (3.51)$$

(3.51) has a soliton solution, that is, a smooth solution that vanishes at infinity. Petviashvili (1975) has found this solution using the stabilizing-multiplier method described in the preceding chapter. It is readily shown that the soliton energy in (3.48) is positive, while the number of particles is proportional to A^{-3}, hence $\partial_A N < 0$, and such a soliton is unstable with respect to spreading in accordance with the stability criterion (3.29).

We proceed to investigate the 1D steady-state solutions of (3.46). To do this, we rewrite it in the form

$$i\partial_t E + \partial_z^2 E + |E|^2 E = \alpha \nabla_\perp^2 \int\!\!\int E dz' dz'', \quad (3.52)$$

where $E = -\partial_z \psi$ is the z component of the electric field.

The 1D periodic solution of (3.52) is

$$E(z, t) = Af(\zeta)\sigma, \quad (3.53)$$

where $\zeta = A(z - 2pt)$, $\sigma = \exp\,[i(\pm A^2 - p^2)t + ipz]$. A and p are constants and σ describes the phase modulation. The conditions for applicability of (3.52) impose some restrictions on A and p, namely, $0 < A \ll 1$ and $|p| \ll 1$ (small amplitude and group velocity). The real-valued function $f(\zeta)$ must, in accordance with (3.52), satisfy the equation

$$d_\zeta^2 f = \pm f - f^3, \quad (3.54)$$

which has periodic solutions of the following two types:

$$f(\zeta) = \left(\frac{2\varkappa^2}{|2\varkappa^2 - 1|}\right)^{1/2} \operatorname{cn}\left(\frac{\zeta}{(|2\varkappa^2 - 1|)^{1/2}}, \varkappa\right), \quad (3.55)$$

$$f(\zeta) = \left(\frac{2}{2 - \varkappa^2}\right)^{1/2} \operatorname{dn}\left(\frac{\zeta}{(2 - \varkappa^2)^{1/2}}, \varkappa\right). \quad (3.56)$$

Here cn and dn are Jacobi's elliptic functions and $\varkappa$ is their modulus. The upper sign in (3.54) gives (3.55), as well as (3.56), when $\varkappa^2 > 0.5$. The lower sign only gives the single periodic solution (3.56), when $\varkappa^2 < 0.5$. The period of (3.55) is $\lambda = 4K(\varkappa)(2\varkappa^2 - 1)^{1/2}$ and that of (3.56) is $\lambda = 2K(\varkappa)(2 - \varkappa^2)^{1/2}$, where $K(\varkappa)$ is the complete elliptic integral. Vedenov and Rudakov (1956) have investigated the stability of periodic waves with constant amplitude $f = 1$. They show that this type of solution is unstable with respect to modulation along the direction of propagation with a growth rate of

$$\gamma = k(2A^2 - k^2)^{1/2} \leq A^2, \quad (3.57)$$

where k is the perturbation wavenumber. Let us examine the steady-state solutions (3.55), (3.56) for stability with respect to infinitesimal perturbations. To do this, we represent E in the form

$$E(z,x,t) = A\left[f(\zeta) + \varphi_1(\zeta, t) + i\varphi_2(\zeta, t)\right]\sigma, \quad (3.58)$$

where the real-valued functions φ_1 and φ_2 are perturbations to the main solution $f(\zeta)$. Substituting (3.58) into (3.52) and linearizing with respect to φ_1 and φ_2, we arrive at a set of two equations with real coefficients

$$\partial_t\varphi_1 = A^2L_2\varphi_2 + \alpha k^2\int\int\varphi_2 d\zeta' d\zeta'', \quad (3.59)$$

where the operator L_2 is defined as

$$L_2 = \left[1 - \partial_\zeta^2 - f^2\right], \quad L_2 f = 0, \quad (3.60)$$

and

$$\partial_t\varphi_2 = -L_1\varphi_1 - \alpha k^2\int\int\varphi_1 d\zeta' d\zeta'', \quad (3.61)$$

with

$$L_1 = 1 - \partial_\zeta^2 - 3f^2 \text{ and } L_1 \partial_\zeta f = 0. \tag{3.62}$$

We first consider the stability with respect to 1D perturbations (with respect to self-modulation). To do this, we set $k = 0$. Choosing a time dependence of the form

$$\varphi_2(\zeta, t) = \psi(\zeta) \exp(A^2 \gamma t), \tag{3.63}$$

we find

$$L_1 L_2 \psi(\zeta) = \gamma^2 \psi(\zeta). \tag{3.64}$$

The problem thus reduces to finding the eigenvalues γ^2 for the ordinary differential equation (3.64) with periodic coefficients subject to the condition that the solution is bounded everywhere (Pavlenko and Petviashvili, 1982). If all the γ^2 for given $\varkappa$ are real and negative, then the solution is stable. Otherwise, it is unstable with a growth rate of $\simeq A^2 \operatorname{Re} \gamma$. We seek a solution to (3.64) in the form of a Floquet solution:

$$\psi(\zeta) = u(\zeta) \exp(ik_0 q \zeta), \quad -1/2 \le q \le 1/2. \tag{3.65}$$

Here $u(\zeta)$ is a periodic function with period $2\pi/k_0$, k_0 is the principal wave number of the periodic function $f^2(\zeta)$, and $k_0 q$ is the characteristic index (quasi-momentum). The stability problem for the periodic wave then reduces to finding the energy of the quasi-particles, γ^2, as a function of $k_0 q$ on a 1D crystal lattice. This problem can be solved algebraically by expanding $u(\zeta)$ in the Fourier series

$$u(\zeta) = \sum_n C_n \exp(ink_0 \zeta). \tag{3.66}$$

Taking the Fourier transformation of (3.64), we find

$$\sum_n L_{mn} C_n = -\gamma^2 C_m, \tag{3.67}$$

where

$$\begin{aligned} L_{mn} = {} & \left[1 + k_0^2 (n + q)^2\right]^2 \delta_{mn} + 3D_2(m - n) \\ & - \left\{4 + \left[3k_0^2 (n + q)^2 + k_0^2 (m + q)^2\right]\right\} D_1(m - n), \end{aligned} \tag{3.68}$$

and

$$D_1(n) = \frac{k_0}{2\pi} \int_0^{\pi/k_0} f^2 \exp(ink_0\zeta)d\zeta,$$

$$D_2(n) = \frac{k_0}{2\pi} \int_0^{2\pi/k_0} f^4(\zeta) \exp(ink_0\zeta)d\zeta.$$

Equations (3.67) are solved by finding the eigenvalues of the infinite matrix (3.68). It can be shown that D_1 and D_2 fall off rapidly with increasing $|m - n|$ for $\varkappa^2 < 0.999$. On the other hand, a numerical study of the behaviour of the eigenvalues with an increasing number of included harmonics shows that, beginning with a 20×20 matrix, the values of the unstable eigenroots for different q and $\varkappa$ are independent of the rank of the matrix. Numerical calculations show that eigenvalues $\gamma^2 > 0$ always exist when $q \neq 0$. This shows that the periodic solutions under consideration are always unstable. However, when $\varkappa \to 1$ the solution (3.64) becomes a soliton, for which the present method is not suitable. Since the transition to a soliton corresponds formally to the vanishing of the quasi-momentum ($k_0 q \to 0$), one can deduce that the soliton is stable from the fact that γ vanishes when $q = 0$. Stability of a 1D soliton can be rigorously proved within the framework of the 1D Schrödinger equation using the ISP method.

Let us examine the stability of the 1D soliton. We proceed from the set of equations (3.59) and (3.61), where f is a soliton solution of (3.54). We assume the time dependence has the form (3.63), as above. The hierarchy of small parameters $\gamma \sim k \gg k^2$ is defined. Accordingly, we set $\varphi_1 = u_0 + u_1 + u_2 + \ldots \varphi_2 = U_0 + V_1 + \ldots$. The zeroth order approximation then gives

$$L_1 u_0 = 0, \quad L_2 V_0 = 0. \tag{3.69}$$

Subsequent approximations applied to (3.59) and (3.61) yield

$$L_1 u_1 = -\gamma V_0, \quad L_2 V_1 = \gamma u_0, \tag{3.70}$$

$$L_1 u_2 = \gamma u_1 + \alpha k^2 \int\int u_0 d\zeta' d\zeta'', \tag{3.71}$$

and

$$L_2 u_2 = -\gamma u_1 - \alpha k_2 \int\int V_0 d\zeta' d\zeta''. \tag{3.72}$$

The solutions of (3.69) are chosen in the form

$$u_0 = \partial_\zeta f, \quad V_0 = 0. \tag{3.73}$$

Substituting (3.73) in (3.70), we find

$$u_1 = 0, \quad V_1 = (\gamma/2)\zeta f. \tag{3.74}$$

From (3.72) we get $V_2 = 0$, and from (3.71) we have

$$L_1 u_2 = -(\gamma^2/2)\,\zeta f + \alpha k^2 \int f d\zeta. \tag{3.75}$$

After multiplying (3.75) by f' and integrating over ζ, on integrating by parts and using the relation $L_1 f' = 0$ we obtain:

$$\gamma^2 = 4\alpha k^2. \tag{3.76}$$

It follows from (3.76) that when $\alpha > 0$, i.e., when $\omega_{Be} < \omega_{pe}$ (weakly magnetized plasma), a 1D soliton is unstable. In a strongly magnetized plasma ($\omega_{pe} < \omega_{Be}$), 1D Langmuir solitons are stable. One-dimensional Langmuir turbulence has been discussed by Rudakov (1973).

3.4. COLLAPSE AND CAVITONS

It has been shown above that no stable steady-state waves can exist in 3D space in a weakly magnetized plasma ($\omega_{pe} > \omega_{Be}$). Zakharov (1972, 1984) has discussed the evolution of a wave packet in such a medium. Few analytical results have been obtained because of the complexity of the problem. However, there are a great many papers in which computer calculations are performed to trace the evolution of a wave packet with different initial conditions. It is beyond doubt at present that a sufficiently compact wave packet, that obeys the nonlinear Schrödinger equation (3.7) or the Zakharov equation (3.36) with bounded integrals of particle number N and Hamiltonian H when $H < 0$, will contract with increasing amplitude. If, on the other hand, $H > H_s > 0$ (H_s is the value of the Hamiltonian for a 3D soliton solution), then the packet will spread. The effect of spontaneous wave packet contraction due to nonlinearity was discovered by Zakharov (1972) and called collapse. This phenomenon can be studied qualitatively in the case of the nonlinear Schrödinger equation by following the analysis of Zakharov (1984).

We define the mean square size of a wave packet as $R = (M/N)^{1/2}$, where

$$M = \int r^2 |\psi|^2 d^3x \quad \text{and} \quad N = \int |\psi|^2 d^3x. \tag{3.77}$$

Here r is the radial coordinate in a spherical coordinate system.

N is an integral of motion according to (3.77). Let us examine the evolution of M. Using the nonlinear Schrödinger equation (3.7), we find

$$\partial_t^2 M = 2mH - 2(m-2)\int |\nabla\psi|^2 d^3x, \tag{3.78}$$

where m is the space dimensionality.

Since H is also an integral of motion from (3.78), we obtain the following estimate for $m \geq 2$

$$\partial_t^2 M \leq 2mH, \tag{3.79}$$

or

$$M \leq mHt^2 + a_1 t + a_2, \tag{3.80}$$

where $a_{1,2}$ are constants of integration.

One can see from this that when $H < 0$, the effective packet size starts to decrease beginning at some instant of time and vanishes during a finite interval of time. Because N and H are conserved, it follows that the packet amplitude tends to infinity. Under real conditions, by this time equation (3.80) is no longer applicable, as effects that were neglected in deriving the nonlinear Schrödinger equation begin to operate. These may be dispersion corrections (terms involving higher-order derivatives) or terms involving higher degrees of nonlinearity. With a suitable sign, both these effects can decelerate collapse, when the size is small and the amplitude is large. The parameters for which collapse ceases can be determined from the Vakhitov–Kolokolov criterion (3.29). Suppose a collapse results in a stable soliton of amplitude A. $\partial_A N < 0$ is always true for Schrödinger's equation in the 3D case. However, Vakhitov and Kolokolov (1973) showed that when the correction terms are incorporated, it may turn out that $\partial_A N > 0$ for a large enough amplitude. The wave packet amplitude for which collapse is retarded can also be determined from (3.29) with the correction terms added. A straightforward treatment of this sort is not possible for Langmuir waves. As numerical experiments demonstrate, although collapse does occur, its evolution follows a different scenario from that for Schrödinger's equation. The principal difference is a tendency for the symmetry to disappear, while the evolution of a packet. In addition, packet contraction occurs so rapidly for large amplitudes that the ion inertia should be taken into account, as it somewhat retards, but does not stop the collapse (Zakharov, 1984).

It can be shown that the most significant correction during the collapse of Langmuir waves is the electron Landau damping. This effect arises when the size approaches the Debye radius. The energy of a collapsing packet is initially transferred to electrons in the tail of the distribution function. As a result, fast electrons appear during collapse. This is one of the indications of collapse in experiments. The question arises whether the Landau damping can retard collapse, as dispersion and nonlinear corrections do. Petviashvili and Tsvelodub (1980) point out that this can take place, provided there is a suitable energy source that is able to compensate the losses due to Landau damping. Likely sources include the parametric instability of an electromagnetic wave during laser heating of a plasma or the beam instability. Zakharov's equation (3.36) must then be sup-plemented by two new terms. (3.36) becomes

$$\nabla^2(i\partial_t\psi + \nabla^2\psi) + \text{div}\,(|\nabla\psi|^2\nabla\psi) = \Gamma\psi - L\,\psi. \qquad (3.81)$$

Here Γ is an operator that describes Langmuir wave pumping in the region of small wavenumbers, L is an operator that incorporates electron Landau damping. Both these operators are given in the linear approximation as their Fourier transforms in wavenumber space. The transform of L is known to be

$$L_k = \text{const}\cdot k^{-3}\exp(-k^{-2}), \qquad (3.82)$$

where k is the wavenumber in units of the inverse Debye radius, while the spectrum of Γ is controlled by the properties of the source through the growth rate in k space. Steady-state localized solutions of (3.81) are called cavitons; they were found numerically by Petviashvili and Tsvelodub (1980) using the stabilizing-multiplier method. A solution is sought in the form $\psi = \varphi(r)\exp(-i\Omega t)$, where Ω is caviton frequency. This is determined during the calculations as an eigenvalue of (3.81) and the solution φ is obtained as a function, that is localized in all directions.

A number of experimental studies have been conducted on the interaction between plasma and electromagnetic waves with a frequency close to the plasma frequency. These revealed clusters of Langmuir waves in which the plasma density was lower. Cheung and Wong (1985) claim that such clusters are due to retardation in the contraction of Langmuir wave packets generated by the parametric instability. The cavitons mentioned above seem to correspond to the clusters found in the experiments of Cheung and Wong (1985).

3.5. MULTISOLITON SOLUTIONS OF THE NONLINEAR SCHRÖDINGER EQUATION

As with the Kadomtsev–Petviashvili equation, dispersion corrections to the nonlinear Schrödinger equation lead to solutions in the form of multisolitons. The solitons that form a Kadomtsev–Petviashvili multisoliton travel with the same speed, while those in a Schrödinger multisoliton have the same propagation velocity and oscillation frequency. We note also that when no corrections have been included in the nonlinear Schrödinger equation, the relevant soliton solutions are comparatively complex in the 2D and 3D cases. These solutions are functions of radius only and can experience numerous changes of sign before vanishing exponentially at infinity.

Gorshkov, Mironov and Sergeev (1983) have investigated the interesting case of a correction to the nonlinear Schrödinger equation consisting of a non-local nonlinearity due to dispersion in the low-frequency (sonic) component and replacement of the Schrödinger equation by a system of equations for the high-frequency and low-frequency components,

$$i\partial_t\psi + \nabla^2\psi + \theta\,\psi = 0, \tag{3.83}$$

$$\theta - \beta\nabla^2\theta = -|\psi|^2. \tag{3.84}$$

This system of equations describes, for example, the self-focusing of electromagnetic waves in plasmas. Here ψ is the dimensionless wave amplitude and θ is the relative temperature perturbation. The same equation describes the interaction of nucleons with the wave function ψ through a meson field with potential θ. The amplitude Φ of the steady-state solution to (3.83), $\psi = \Phi(r)\exp(iEt)$, satisfies the equation

$$\nabla^2\Phi + (\theta - E)\Phi = 0. \tag{3.85}$$

Soliton solutions of (3.84) and (3.85) in the 2D and 3D cases are known to be unstable when $\beta = 0$. Using the Vakhitov–Kolokolov criterion (3.29), one can show that, when $\beta > 0$, solitons with a large enough value of E become stable, multisoliton solutions also appear in this case.

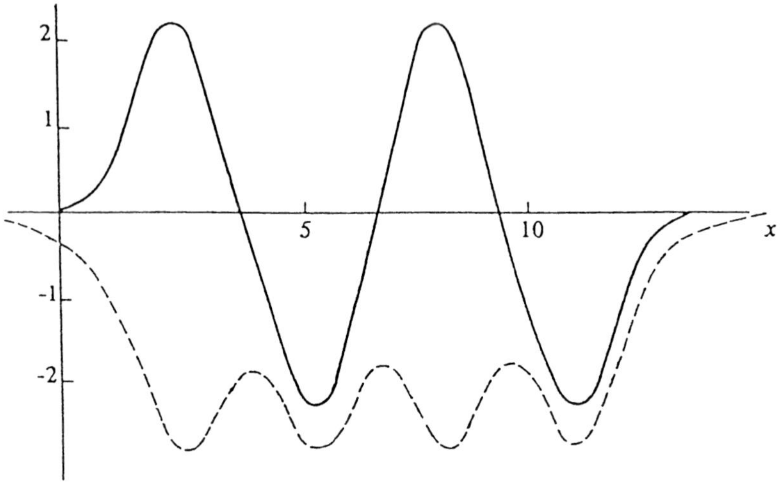

Figure 3.2. The one-dimensional quartet soliton solution of equations (3.84) and (3.85): a cross-section of Φ (solid line), and cross-section of θ (dashed line).

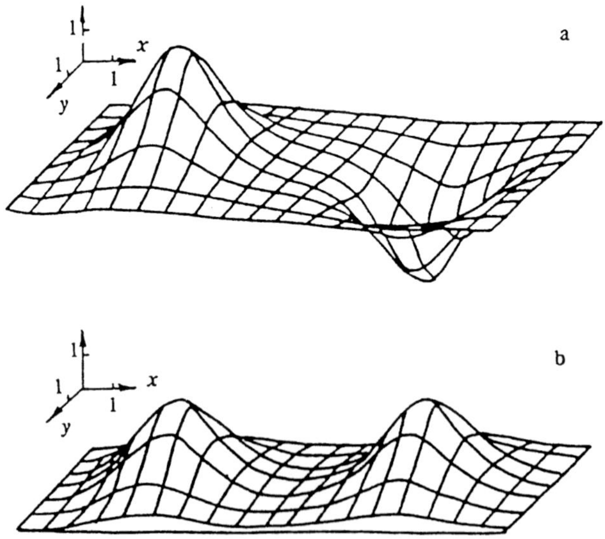

Figure 3.3. The 2D bisoliton solution of (3.84), (3.85): contours (a) Φ and (b) θ.

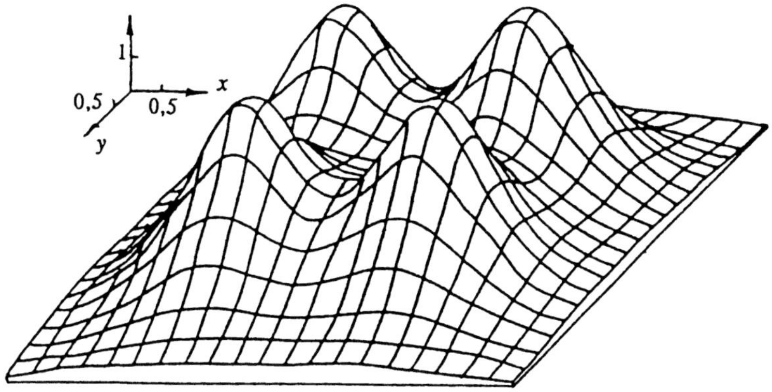

Figure 3.4. Contours of θ for the 2D quartet soliton solution of (3.84) and (3.85).

According to Gorshkov, Mironov and Sergeev (1983), multisoliton solutions of (3.84) and (3.85) occur when $\beta E > 1$. Figure 3.2 shows a 1D solution of this system for $\beta E = 4$, Figure 3.3 contains a 2D solution of this system in the form of a bisoliton, and a quartet soliton is shown in Figure 3.4. In the case of nuclear matter, the 3D equations (3.83) and (3.84) also have a multisoliton solution, corresponding to a nucleus composed of several nucleons. The system of equations should then be supplemented by a normalization condition for the number of nucleons and by Pauli's principle. The first term in (3.84) can be disregarded for large energies ($\beta E \gg 1$); combined with the normalization condition, this yields an equation describing the Coulomb self-interaction of the wave function. In this case equation (3.84) becomes Poisson's equation.

4. Diamagnetism of cyclotron plasma waves and cyclotron solitons

4.1. NONLINEAR VARIATION OF THE MAGNETIC FIELD

Both electron and ion cyclotron waves are important as an effective source of auxiliary heating in plasmas. When high-power radiation is applied to a plasma (as much as a few MW), it produces nonlinear effects that affect the propagation and absorption of beams of waves. Experimental observations of cyclotron oscillations (TFR Group, 1975) show that they play an important role in the behaviour of laboratory plasmas. Cyclotron waves also occur widely in the Earth's magnetosphere (Russel *et al.*, 1970; Gurnett, 1974; Bergmann, 1984). In particular, intensive emission of electromagnetic waves in the electron cyclotron frequency range has been recorded in the auroral region; this is the Earth's kilometre wavelength radiation (Gurnett, 1974; Alexander and Kaiser, 1977; Cole and Pokhotelov, 1980). In a number of cases, cyclotron waves are observed to have large amplitudes and to be localized in space (Mozer *et al.*, 1977), possibly as a consequence of nonlinearity.

The first step in studying highly nonlinear cyclotron waves is to derive simplified equations that include only the main linear and nonlinear effects. In Langmuir waves, because of the dominant dependence of Langmuir frequency on plasma density, the principal mechanism responsible for the highly nonlinear effects discussed above (such as wave collapse or the creation of solitons) is the formation of density cavities in the localization region of a wave packet. The frequency of cyclotron waves is largely controlled by the ambient magnetic field. For this reason, as Petviashvili (1976b) has shown, the principal nonlinear mechanism that determines the behaviour of cyclotron waves is the formation of cavities in a ambient magnetic field within the region of localization of cyclotron waves.

The reduction in the magnetic field in a wave packet localization region which was referred to as high-frequency (HF) diamagnetism by Petviashvili (1976b), had been discussed previously by Landau and Lifshitz (1963), Pitaevsky (1961), and Washimi and Karpman (1976). Petviashvili and Pokhotelov (1977) discovered the self-focusing effect of HF diamagnetism on Alfvén waves.

HF diamagnetism usually arises because of a diamagnetic current which develops in a plasma transverse to the constant magnetic field in the wave packet localization region owing to the high frequency pressure (ponderomotive force) acting on the plasma. When Pitaevsky (1961) derived an expression for diamagnetic lowering of the magnetic field in a packet, δB, he neglected the effect of spatial dispersion, that is, the non-local interaction between an electric field and matter associated with thermal motion. The expression for δB in a cold plasma obtained by Pitaevsky (1961) has the form

$$\delta \mathbf{B} = \frac{1}{4} \frac{\partial \varepsilon_{ik}}{\partial \mathbf{B}_0} E^*_{0\,i} E_{ik}, \tag{4.1}$$

where E_0 is the complex amplitude of oscillatory electric field and ε_{ik} is the dielectric permeability tensor of a cold plasma.

Nekrasov and Petviashvili (1979) used a straightforward procedure to obtain the nonlinear variation in the magnetic field δB including spatial dispersion and a new effect, the nonlinearity in the cyclotron resonance for electrostatic cyclotron waves (Bernstein modes). The general expression is

$$\begin{aligned}
\delta B = & -\sum_{k',k''} \langle \varphi_{\mathbf{k}'\omega'} \varphi_{\mathbf{k}''\omega''} \rangle \exp[i(\mathbf{k}' + \mathbf{k}'')\mathbf{r} - i(\omega' + \omega'')t] \\
& \times \sum_j \frac{\omega_{pj}^2 \omega_{Bj}}{2B_0} \sum_n |\mathbf{k}_\perp' + \mathbf{k}''|^{-1} \left[\frac{\partial}{\partial |\mathbf{k}_\perp' + \mathbf{k}_\perp''|} |\mathbf{k}_\perp' + \mathbf{k}_\perp''| S_n^{(1)} \right. \\
& \left. + i \frac{k_\perp' k_\perp''}{n \omega_{Bj}^2} S_n^{(2)} \sin(\alpha_{\mathbf{k}'} - \alpha_{-\mathbf{k}''}) \right] \frac{(n\omega_{Bj})^2}{(\omega' - n\omega_{Bj})(\omega'' + n\omega_{Bj})} \\
& \times \exp\left[in(\alpha_{\mathbf{k}'} - \alpha_{-\mathbf{k}''}) \right],
\end{aligned} \tag{4.2}$$

where

$$S_n^{(p)} \equiv 2\pi \int_0^\infty v_\perp^{2(p-1)} J_{1,\mathbf{k}'+\mathbf{k}''} J_{n,\mathbf{k}'} J_{n,\mathbf{k}''} \frac{1}{v_\perp} \frac{\partial f_{oj}}{\partial v_\perp} dv_\perp \tag{4.3}$$

Here $\varphi_{\mathbf{k}\omega}$ is the Fourier component of the oscillating potential, ω_{pj} is the Langmuir frequency for particles of species j, ω_{Bj} is the cyclotron frequency, $\alpha_\mathbf{k}$ is the azimuthal angle in the wave vector space on a plane perpendicular to the ambient magnetic field B_0 aligned along the z-axis, $J_{n,\mathbf{k}} \equiv J_n(k_\perp v_\perp / \omega_{Bj})$ is the n-th Bessel function, f_{oj} is the unperturbed function of particle distribution, and angular brackets $\langle ... \rangle$ denote averaging over the HF oscillations. It was assumed in the derivation of (4.2) that $\omega - n\omega_{Bj} \gg k_z v_{T\parallel j}$, where $v_{T\parallel j}$ is longitudinal thermal particle velocity. The expression (4.2) is valid both for random and regular fields with an arbitrary transverse wavelength. We begin by examining the nonlinear variation in the magnetic field owing to turbulence. Averaging (4.2) over the random phases, we obtain

$$\delta B = \frac{1}{2} \sum_{\mathbf{k},\omega} k^2 |\varphi_{\mathbf{k}\omega}|^2 \frac{\partial \varepsilon(\mathbf{k},\omega)}{\partial B_0}, \tag{4.4}$$

where

$$\varepsilon(\mathbf{k},\omega) = 1 + \sum_j \frac{\omega_{pj}^2}{k^2} \sum_n 2\pi \int_0^\infty J_{n,\mathbf{k}}^2 \frac{n\omega_{Bj}}{\omega - n\omega_{Bj}} \frac{\partial f_{oj}}{\partial v_\perp} dv_\perp \tag{4.5}$$

is the dielectric permeability for electrostatic waves. Note that the differentiation with respect to B_0 in (4.4) is to be performed only for the cyclotron frequency in the resonant denominator in (4.5). Formula (4.4) is similar to (4.1). In the cold plasma approximation, where $n = \pm 1$ and $k_\perp r_{Bj} \ll 1$, r_{Bj} denotes the Larmor radius for particles of species j, for sufficiently narrow-band wave spectrum, we obtain

$$\delta B = \frac{1}{4} \frac{\partial \varepsilon}{\partial B_0} |\nabla \psi|^2, \tag{4.6}$$

where ψ is the amplitude of the oscillations in the potential in coordinate space.

For regular oscillations, the expression (4.2) takes a form analogous to (4.4) only when the conditions $|\mathbf{k}_\perp' + \mathbf{k}_\perp''| r_{Bj} \ll 1$ and $k_\perp^2 r_{Bj}^2 \Delta\alpha_\mathbf{k} \ll 1$ are satisfied, where $\Delta\alpha_\mathbf{k}$ is the angular spread of the wave vectors. The other cases are not all covered by a formula of the type (4.6). Nekrasov and Petviashvili (1979) have obtained, in particular, an expression for the nonlinear correction to the magnetic field in the case of short-wave cyclotron oscillations with $k_\perp r_{Bj} \gg 1$ that possess cylindrical symmetry with respect to an axis of symmetry along the ambient magnetic field. For this type of particle it has the form

$$\delta B = \frac{\pi}{2B_0 r_{Bj}^4} \sum_{\omega',\omega''} \frac{n^2\omega_{pj}^2}{(\omega' - n\,\omega_{Bj})(\omega'' + n\,\omega_{Bj})} \int_0^\infty \int_0^\infty \frac{d\mathbf{k}_\perp' d\mathbf{k}_\perp''}{(2\,\pi)^4}$$

$$\times \langle \varphi_{\mathbf{k}_\perp'\omega'}\,\varphi_{\mathbf{k}_\perp''\omega''}\rangle J_0(|\mathbf{k}_\perp' - \mathbf{k}_\perp''|\mathbf{r}_\perp) I_1\left[2(\mathbf{k}_\perp' - \mathbf{k}_\perp'')^2 r_{Bj}^2\right]$$

$$\times \exp\left[-2(\mathbf{k}_\perp' - \mathbf{k}_\perp'')^2 r_{Bj}^2\right], \tag{4.7}$$

where $\mathbf{r}_\perp$ is the distance from the axis of symmetry, I_1 is a Bessel function of an imaginary argument, $r_{Bj} = (T_{\perp j}/2m_j\omega_{Bj}^2)^{1/2}$, and $n > 0$. The distribution function f_{oj} is assumed to be Maxwellian.

4.2. MODEL EQUATION FOR LONG WAVELENGTH CYCLOTRON OSCILLATIONS

The nonlinear equation for cyclotron waves of finite amplitude is derived, as in the case of Langmuir oscillations, by solving the dispersion relation $\varepsilon(\mathbf{k}, \omega) = 0$ for cyclotron waves $\omega = n\,\omega_{Bj}(1 + R_j)$, where $R_j \ll 1$, and including a nonlinear correction to the cyclotron frequency. If the electric potential is represented as an oscillating function with frequency $n\,\omega_{Bj}^0$ and envelope ψ:

$$\varphi(\mathbf{r}, t) = \frac{1}{2}\left\{\psi(\mathbf{r}, t) \exp\left(-i\,n\,\omega_{Bj}^0 t\right) + c.c.\right\}, \tag{4.8}$$

then the equation for ψ in the parabolic approximation for $k_\perp r_{Bj} \ll 1$ becomes

$$\nabla^2\left(\frac{i}{n\omega_{Bj}^0}\,\partial_t\psi - Q_{n,j}\psi\right) = \operatorname{div}(h\nabla\psi),$$

$$h = -a_{nj}|\chi|^2, \tag{4.9}$$

$$\frac{i}{n\,\omega_{Bj}^{0}}\,\partial_t\chi - h\chi = P_{n,j}\,\psi,$$

where $h = \delta B/B_0$ is the relative depth of the magnetic well, $a_{n,j} = (4n!r_{Dj}^2 B_0^2)^{-1}$, $r_{Dj} = (T_{\perp j}/m_j)^{1/2}\omega_{pj}^{-1}$ is the Debye radius, and $P_{n,j} \equiv r_{Bj}^{\,n}(\partial_x + i\partial_y)^n$. The nonlinear term in the equation for χ is the result of including nonlinear variation of the cyclotron frequency in the resonance denominators of (4.2).

Equations (4.9) are valid for both electron and ion cyclotron oscillations. In the former case $Q_{n,e} = \alpha_{n,e}|P_{n-1,e}|^2$, where

$$\alpha_{n,e} = \frac{(-1)^{n-1}}{2n!}\,\frac{(n^2-1)\omega_{pe}^2}{(n^2-1)(\omega_{Be}^0)^2 - \omega_{pe}^2} \tag{4.10}$$

(ω_{pe}^2 should be omitted from the denominator when $n = 1$). In the latter case, when $n\,\omega_{Bi} \ll k_z v_{T\|e}$, we have $Q_{n,i} = \alpha_{ni}|P_{n,i}|^2$, where

$$\alpha_{n,i} = \frac{(-1)^n}{n!}\,\frac{T_{\|e}}{T_{\perp i}}. \tag{4.11}$$

Borodachev and Nekrasov (1984) have used the system of equations (4.9) for electron cyclotron oscillations in the second harmonic ($n = 2$) to examine the modulational instability and self-focusing of a monochromatic travelling wave. The instability growth rate γ increases with increasing amplitude. For a sufficiently strong wave $E_0/B_0 > (k_0\,r_{Be})^2$, the growth rate for the modulational instability ($\mathbf{k}_0 \parallel \vec{\varkappa}$) is the greatest when $\varkappa \gg k_0$. Here E_0 is the amplitude of the electric field of the wave, $\mathbf{k}_0$ is its wave vector, and $\varkappa$ the perturbation wave vector. When $\omega_{pe} \simeq \omega_{Be}^0$, to within an order of magnitude we have

$$\gamma \simeq (k_0 r_{Be})^{-2/3}(E_0/B_0)^{4/3}\omega_{Be}^0. \tag{4.12}$$

The maximum of the self-focusing growth rate ($\mathbf{k}_0 \perp \varkappa$) is attained at $\varkappa \sim k_0$, and when $\omega_{pe} \sim \omega_{Be}^0$ for a strong wave it is equal, within an order of magnitude, to

$$\gamma \simeq (E_0/B_0)\,\omega_{Be}^0. \tag{4.13}$$

Note that the instability has no threshold when $(\mathrm{k}_0 \times \varkappa)_z \neq 0$.

Equations (4.9) admit of steady-state solutions, that either are 1D or are functions of the radius in cylindrical coordinates. We examine the case of first harmonic $(n = 1)$ ion cyclotron waves as the simplest and most important (Nekrasov and Petviashvili, 1979). We seek a solution in the form

$$\psi = \frac{2^{1/2}}{g_i^{1/2}\mu^2} A^3 B_0 f(\rho) \exp\,(i\, A^2 \omega_{Bi}^0 t), \quad h = -A^2 F(\rho), \tag{4.14}$$

where

$$g_i = \frac{\omega_{pi}^2}{(\omega_{Bi}^0)^2}, \quad \rho = \mu r_\perp, \quad \mu = \left(\frac{T_{\perp i}}{T_{\parallel e}}\right)^{1/2} A r_{Bi}^{-1}, \quad A^2 \ll 1.$$

Substituting (4.14) into the system (4.9), we get

$$\mathrm{div}\,(1 - \nabla^2)\mathrm{E} = \mathrm{div}\, F\,\mathrm{E}, \tag{4.15}$$

$$F(1 - F)^2 = D^2 \equiv |E_x + iE_y|^2, \tag{4.16}$$

where $\mathrm{E} = \nabla f$, and f is the dimensionless potential amplitude.

In accordance with (4.14), F determines the spatial variation of the magnetic well depth. From (4.16) it follows that the dependence of the well depth on the amplitude D of the electric field is not a unique function. Under real conditions we have a dependence in which $dF/dD > 0$. This is represented by the solid line in Figure 4.1. The portion aa' does not

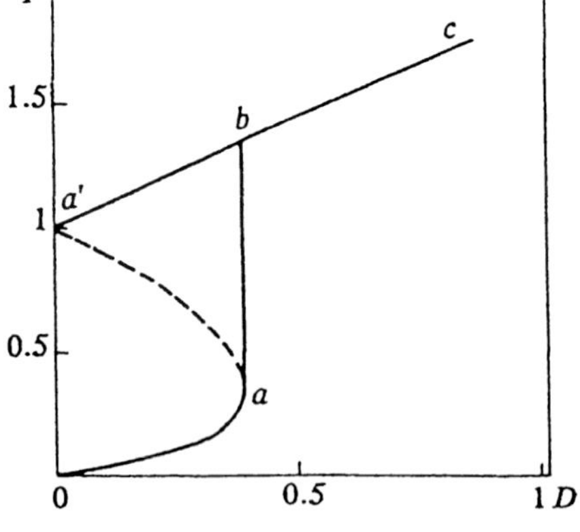

Figure 4.1. The magnetic well depth as a function of the electric field amplitude in the steady-state case (in dimensionless units)

actually exist, because in it the depth of the magnetic well decreases with increasing D. The depth of the magnetic well varies discontinuously along the portion ab in the vicinity of the exact cyclotron resonance ($F = 1$). The point b is thus a branching point.

If all quantities in (4.15) are functions of the dimensionless coordinate ξ only ($\xi \equiv \mu x$), then E lies in the ξ direction. The solution of (4.15) and (4.16) for this case is plotted in Figure 4.2. When $\xi < 10$, we have a periodic solution in which the curve $a'bc$ (Figure 4.1) is realized. At $\xi = 10.7$ the depth of the well drops toward point a at point b, and the solution approaches zero as ξ increases.

When f in (4.14) depends on the radius ρ ($\rho = \mu r_\perp$) alone, E is directed along the radius in the cylindrical coordinate system. In that case (4.15) and (4.16) have the solution shown in Figure 4.3.

One can see from the figures that the magnetic field is always discontinuous at the boundaries of localized steady-state solutions, and is the HF pressure (because the latter depends on the magnetic field resonantly). A tangential discontinuity in the current layer is realized, which is analogous to a well-known solution for MHD equations (Landau and Lifshitz, 1963). The amplitude of the jump in the magnetic field is given by (4.16) and is shown in Figure 4.1. The total pressure remains continuous, however, owing to the discontinuity in the HF pressure. If the external electric field E_{ext} is assumed to be the standing wave

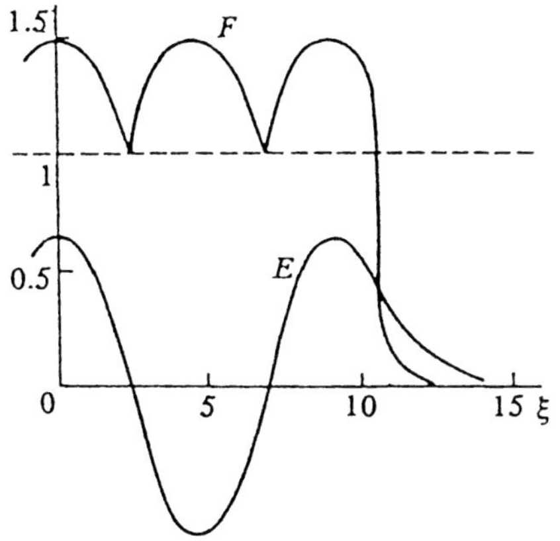

Figure 4.2. The one-dimensional steady-state solution for ion cyclotron waves at the first harmonic. One can see that cyclotron waves can give rise to a periodic structure in the constant magnetic field in the form of domains (curve F, in dimensionless units).

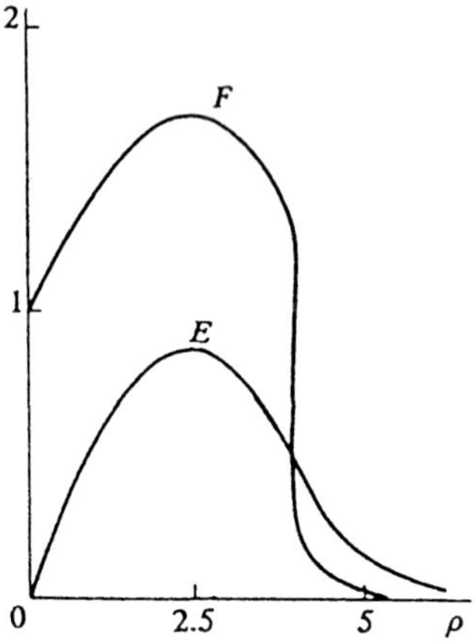

Figure 4.3. Solution of the ion cyclotron wave equations as an axisymmetrical soliton extended along the magnetic field (in dimensionless units).

$$E_{ext} = E^{0}_{ext} \cos\left[(2\omega^{0}_{Be} + \delta\omega_0)t\right] \cos(k_0 x), \tag{4.17}$$

then in the steady state the system (4.9) can yield an equation for the qualitative dependence of the typical depth of the magnetic well $|h_0|$ on the amplitude of the external field:

$$|h_0|(\Delta + |h_0|)^2 = D_0^2, \tag{4.18}$$

where

$$\Delta = \Omega_0 + \alpha_{2,e} k_0^2 r_{Be}^2, \quad D_0 = 2^{-3/2} k_0 r_{Be}^2 r_{De}^{-1} \frac{E^{0}_{ext}}{B_0},$$

$$\Omega_0 = \frac{\delta\omega_0}{2\omega^{0}_{Be}}, \quad \alpha_{2,e} = -\frac{3}{4} \frac{\omega_{pe}^2}{3(\omega^{0}_{Be})^2 - \omega_{pe}^2}.$$

When $\Delta = 0$, the frequency of the external field coincides with that of the eigenmodes.

The numerical simulations were designed to test two theoretical relations derived from (4.18):

$|h_0| = f_1(\Delta)$ when E^{0}_{ext} = constant and $|h_0| = f_2(E^{0}_{ext})$ when $\Delta = 0$. The values of $|h_0|$ for given fields taken from the simulation were compared with the theoretical values. Figures 4.4 and 4.5 show the

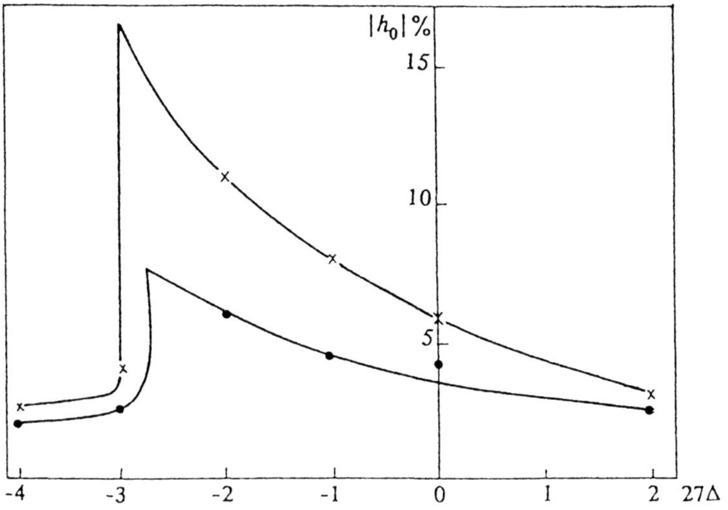

Figure 4.4. The relative depth of the magnetic well $|h_0|$ as a function of Δ for $E^0_{ext}/B_0 = 0.25$ ($\omega_{Pe} = \omega^0_{Be}$, $k_0^2 r_{Be}^2 = 0.2$): x — theory, • — computation.

respective curves. One can see that the theory and numerical simulation are in good agreement. The discontinuity in Figure 4.4 is due to the single-valued dependence of $|h_0|$ on Δ.

Figure 4.6 (curve a) shows a time-averaged profile of the magnetic field for a pump wave of the form $E_{ext} \sim \cos k_0 x$, where $k_0 = 2\pi/L_0$ (L_0 is the length of the system). Two wells are formed. The same result follows from (4.9).

Figure 4.6 (curves b,c) shows time-averaged profiles of the magnetic field for two values of the pump wave amplitude, $E^0_{ext}/B_0 = 0.1$ and $E^0_{ext}/B_0 = 0.175$, respectively for the case $\Delta = 0$. The former value has a profile with two wells, the latter, with three wells. Curve d (Figure 4.6) shows a time-averaged profile of the magnetic field with parameters derived from the theory assigned so as to create four wells. Numerical modelling of magnetoactive plasma thus confirms the existence of the diamagnetic effect.

The diamagnetic effect at the second harmonic of the electron cyclotron frequency has been observed experimentally by Zaleskii *et al.* (1982).

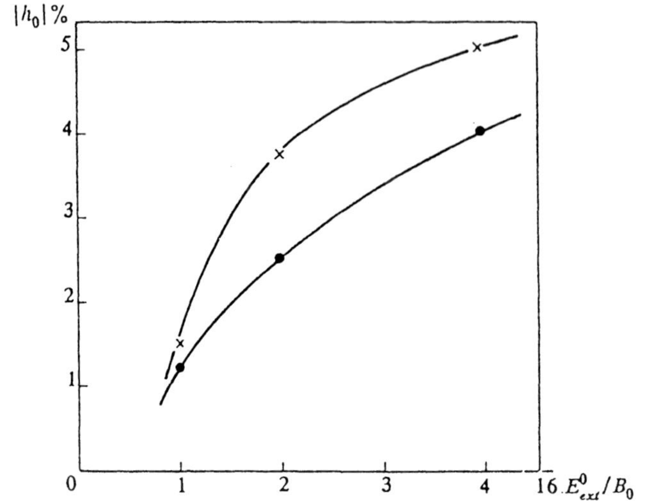

Figure 4.5. The relative depth of the magnetic well $|h_0|$ as a function of E^0_{ext} for $\Delta = 0$ ($\omega_{Pe} = \omega^0_{Be}$, $k_0^2 r_{Be}^2 = 0.2$): x — theory, • — computation.

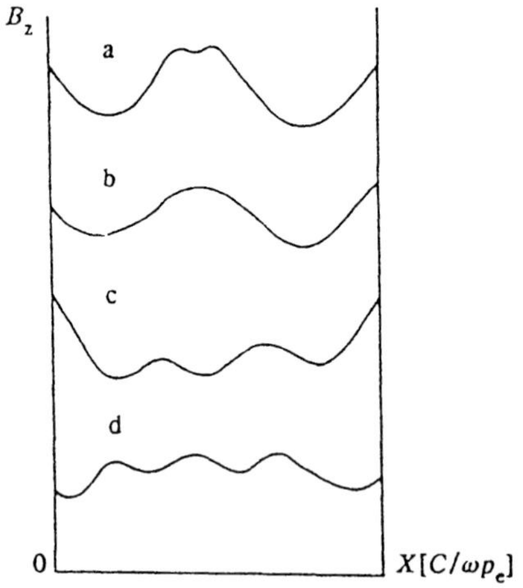

Figure 4.6. Time-averaged profile of the magnetic field $B_z(x)$ $(0 \leq x \leq L_0)$ for (a) standing pump wave with $E^0_{ext}/B_0 - 0.25$ $(k_0 L_0 = 2\pi)$, (b) travelling pump wave with $E^0_{ext}/B_0 - 0.1$ $(k_0 L_0 - 2\pi)$, (c) travelling pump wave with $E^0_{ext}/B_0 - 0.175$ $(k_0 L_0 = 2\pi)$, (d) travelling pump wave with $E^0_{ext}/B_0 - 0.06$ $(k_0 L_0 = \pi/2)$.

4.3. MODEL EQUATIONS FOR SHORT WAVE-LENGTH CYCLOTRON OSCILLATIONS

Consider the case in which the typical wavelength of electrostatic (potential) oscillations transverse to a magnetic field is less than the Larmor radius: $k_\perp \cdot r_{Bj} \gg 1$. We assume at the same time that the amplitude of oscillations varies little over distances on the order of the Larmor radius; that is, we consider a wave packet that is narrow in **k** space (quasimonochromatic travelling wave) under the conditions $|\mathbf{k}_\perp' + \mathbf{k}_\perp''| \cdot r_{Bj} \ll 1$. From these inequalities it follows that the angular spread $\Delta\alpha_{\mathbf{k}}$ and the spread in wave number $\Delta \mathbf{k}_\perp$ are small, i.e. the angle $\Delta\alpha_{\mathbf{k}} \ll (k_\perp r_{Bj})^{-1}$ and $(\Delta k_\perp r_{Bj})^2 \ll 1$.

We now introduce the functions $\chi(\mathbf{r}, t)$ and $\mu(\mathbf{r}, t)$ which obey:

$$\vec{\chi}(\mathbf{r}, t) = -\nabla\mu(\mathbf{r}, t) = \sum_\omega \frac{n\,\omega_{Bj}}{\omega - n\,\omega_{Bj}} \mathrm{E}_\omega e^{-i\omega t}, \tag{4.19}$$

where $\mathrm{E}_\omega = -\nabla\varphi_\omega$. Then, using the asymptotic behauiour of Bessel functions under these conditions for oscillations near the n-th harmonic of the cyclotron frequency ω_{Bj} in the case of very narrow wave packets, when $\Delta\,\alpha_{\mathbf{k}} \ll (k_\perp r_{Bj})^{-2}$, we obtain the following expression for δB (Nekrasov and Feygin, 1985) from (4.2):

$$\delta B = -\frac{2\,\omega_{pj}^2\,\omega_{Bj}^0}{\sqrt{\pi}\,B_0 v_{T\perp j}^3\,k_{\perp 0}}\,|\mu|^2, \tag{4.20}$$

where $v_{T\perp j} = (2T_{\perp j}/m_j)^{1/2}$ and $k_{\perp 0} \simeq k_\perp' \simeq k_\perp''$. As before f_{0j} is assumed to be a Maxwellian distribution function.

If the angular spread $\Delta\,\alpha_{\mathbf{k}}$ is small, but finite, i.e. $(k_\perp r_{Bj})^{-2} \ll \Delta\alpha_{\mathbf{k}} \ll (k_\perp r_{Bj})^{-1}$, then we have

$$\delta B = i\,\frac{\omega_{pj}^2}{2\sqrt{\pi}\,B_0 k_{\perp 0} v_{T\perp j} n\,\omega_{Bj}^0}\,(\chi \times \chi^*)_z. \tag{4.21}$$

Note that (4.20) can be written in the form

$$\delta B = \frac{1}{4} \frac{\partial \varepsilon}{\partial B_0} |\partial_x \psi + i \partial_y \psi|^2.$$

The variation of the magnetic field is thus associated with the HF pressure in this case. Formula (4.21) differs fundamentally from (4.20). In this case, the mechanism for formation of the nonlinear magnetic field does not reduce to the action of the HF pressure. It is analogous to the mechanism for the excitation of a magnetic field by Langmuir oscillations in plasmas without an external magnetic field (Bel'kov and Tsytovich, 1979) and is associated with an inhomogeneity in the amplitude and phase of the wave potential, which makes different components of the electric field have different phases. In particular, (4.21) is nonzero for rotating fields (Landau and Lifshitz, 1963). Note that δB may have different signs in (4.21).

Nekrasov and Feygin (1985) have derived equations for the evolution of the amplitude $\psi(\mathbf{r}, t)$ of a packet of short wavelength cyclotron waves,

$$\varphi(\mathbf{r}, t) = \frac{1}{2} \left\{ \psi(\mathbf{r}, t) \exp\left(-i\omega_0 t + i\mathbf{k}_{\perp 0} \cdot \mathbf{r}_{\perp}\right) + c.c. \right\},$$

where $\mathbf{k}_{\perp 0}$ and ω_0 are the carrier wave vector and packet frequency. These equations were derived using (4.19)–(4.21) and the dispersion relation for small k_z; in the coordinate system moving at the group velocity, they have the form

$$\frac{i}{n\,\omega^0_{Bj}} \partial_t \psi - \alpha \nabla^2_{\perp} \psi + \beta \partial^2_z \psi = -d_1 |\psi|^2 \psi \tag{4.22}$$

for $\Delta\, \alpha_{\mathbf{k}} \ll (k_{\perp 0} r_{Bj})^{-2}$ and

$$\frac{i}{n\,\omega^0_{Bj}} \partial_t \psi - \alpha \nabla^2_{\perp} \psi + \beta \partial^2_z \psi = -d_2 (\mathbf{k}_{\perp 0} \times \nabla_{\perp} |\psi|^2)_z \psi \tag{4.23}$$

for $(k_{\perp 0} r_{Bj})^{-2} \ll \Delta\, \alpha_{\mathbf{k}} \ll (k_{\perp 0} r_{Bj})^{-1}$. Here

$$\alpha = \frac{3}{2} \frac{\delta}{k^2_{\perp 0}}, \quad \beta = \frac{1}{2n^2} \frac{T_{\parallel j}}{T_{\perp j}} \frac{r^2_{Bj}}{\delta}, \quad \delta = \frac{2}{\sqrt{\pi}} \frac{\omega^2_{pj}}{(\omega^0_{Bj})^2 k^3_{\perp 0} r^3_{Bj}},$$

$$d_1 \equiv \frac{\sqrt{\pi}}{8} \frac{(\omega_{Bj}^0)^2}{\omega_{pj}^2} \frac{k_{\perp 0}^5 r_{Bj}^3}{B_o^2}, \quad d_2 \equiv \frac{\sqrt{\pi}}{32n} \frac{(\omega_{Bj}^0)^2}{\omega_{pj}^2} \frac{k_{\perp 0}^5 r_{Bj}^5}{B_0^2}.$$

The maximum growth rate of the instability of a monochromatic wave given by (4.22) is (Nekrasov and Feygin, 1985)

$$\gamma \simeq \frac{\sqrt{\pi}}{8} \frac{(\omega_{Bj}^0)^2}{\omega_{pj}^2} (k_{\perp 0} r_{Bj})^3 \frac{E_0^2}{B_0^2} n\, \omega_{Bj}^0 . \tag{4.24}$$

The self-focusing growth rate given by (4.23) increases with the wave vector of the modulation, $\varkappa(\varkappa \perp \mathbf{k}_{\perp 0})$, and the instability has no threshold (as is the case for $k_\perp r_{Bj} \ll 1$). When the dispersion correction to the frequency owing to the modulation $\varkappa$ alone is greater than the nonlinear correction, we have

$$\gamma \simeq \frac{1}{10} \frac{(\omega_{Bj}^0)^2}{\omega_{pj}^2} \frac{\varkappa}{k_{\perp 0}} (k_{\perp 0} r_{Bj})^5 \frac{E_0^2}{B_0^2} n\, \omega_{Bj}^0 \tag{4.25}$$

Nekrasov and Feygin (1985) have discussed whether the steady-state solutions (solitons) described by (4.22) and (4.23) can exist. In particular, it follows from (4.22) that the envelope soliton for very narrow wave packets is bounded only along the magnetic field (the z-axis). The characteristic length L_z of a soliton along the z-axis is equal to

$$L_z \simeq A^{-1} \frac{\omega_{Bj}^0}{\omega_{pj}} \left(\frac{T_{\parallel j}}{T_{\perp j}} \right)^{1/2} (k_{\perp 0} r_{Bj})^{3/2} r_{Bj} , \tag{4.26}$$

where $A \ll 1$ is the dimensionless amplitude.

4.4. SELF-FOCUSING AND THREE-DIMENSIONAL LOCALIZATION OF A CYCLOTRON WAVE TRAVELLING ALONG A MAGNETIC FIELD

The preceding sections were concerned with the effect of nonlinear changes in the magnetic field due to electrostatic cyclotron waves. We now examine a similar effect for electromagnetic cyclotron waves. We begin by considering waves travelling along the magnetic field (Nekrasov and Petviashvili, 1981).

We know that when a beam of electromagnetic waves propagates, a nonlinear variation in the refractive index N with the appropriate sign of the spatial dispersion leads to self-focusing (Askarian, 1962).

Litvak's review (1986) discusses self-focusing of electromagnetic waves in plasmas owing to the change in N caused by a reduction in the density within the beam under the action of the HF pressure.

When an extraordinary wave propagates along a magnetic field in a plasma, the index of refraction depends significantly on the magnetic field. In this case, a nonlinear variation in N also occurs because of the reduced magnetic field within the wave packet. For sufficiently short wave trains, diamagnetism may prove to be the main nonlinear effect influencing the propagation of the extraordinary wave.

The dispersion relation for an extraordinary wave of frequency ω propagating at a small angle to an ambient magnetic field $\mathbf{B}_0$ directed along the z-axis is derived from the expression for $N^2 = k^2c^2/\omega^2$ in a cold plasma (Litvak, 1986):

$$N^2 = 1 - \frac{\omega_{pe}^2}{\omega(\omega - \omega_{Be})} + \frac{\omega_{pe}^2 \omega_{Be}}{2(\omega^2 - \omega_{pe}^2)(\omega - \omega_{Be})} N_\perp^2 \, ; \tag{4.27}$$

$$N_\perp^2 \equiv \frac{k_\perp^2 c^2}{\omega^2} .$$

It was assumed in the derivation of (4.27) that

$$1 \gg \frac{\omega + \omega_{Be}}{2\,\omega} \frac{k_\perp^2 c^2}{|\omega^2 - \omega_{pe}^2|} \quad \text{and} \quad \omega - \omega_{Be} \gg k_z v_{Te} .$$

(Note that when $\omega \simeq \omega_{pe}$, we get $k^2c^2 \simeq \omega^2$.) Let us assume that ω is close to but smaller than ω_{Be}, i.e. $|\omega - \omega_{Be}| < \omega < \omega_{Be}$. On the right-hand side of (4.27) let us take into account slow nonlinear variations in the density and magnetic field by writing $n = n_0 + \delta n$, $B = B_0 + \delta B$. Then, for a steady-state travelling wave (the condition for steady-state behaviour will be given below) with an x-component of the electric field given by

$$E_x = \frac{1}{2} E_\perp(r_\perp, z) \exp(-i\omega t + ik_{0z} z) + c.c.$$

(the phase of E_y is shifted by $\pi/2$ relative to E_x) and a slowly varying amplitude $E_\perp$ along the z-axis:

$\partial_z \ll k_{0z}$, where $k_{0z} = (\omega/c)(1 + a)^{1/2}$, $a = -(\omega_{pe}^0)^2/\omega(\omega - \omega_{Be}^0)$,

ω_{pe}^0 and ω_{Be}^0 are the unperturbed plasma and cyclotron frequencies, with the aid of (4.27) we obtain the following equation

$$i\partial_\xi E_\perp + \frac{\sigma}{1-F}\nabla_\eta^2 E_\perp + \frac{\alpha}{1-F}\left(F + \frac{\delta n}{n_0}\right)E_\perp = 0. \tag{4.28}$$

Here

$$F = \frac{\omega_{Be}^0}{\omega - \omega_{Be}^0}\frac{\delta B}{B_0}, \quad \alpha = \frac{a}{2(1+a)^{1/2}}, \quad \xi = \frac{\omega z}{c},$$

$$\eta = q\, r_\perp, \quad q = \left[\frac{4(1+a)^{1/2}}{a}\frac{|\omega^2 - (\omega_{pe}^0)^2|}{c^2}\frac{\omega}{\omega_{Be}^0}\right]^{1/2},$$

$$\sigma = \operatorname{sign}[\omega^2 - (\omega_{pe}^0)^2].$$

It is clear from (4.28) that both the numerator and the denominator contain nonlinearities. When $F = 1$, the frequency ω coincides with the local cyclotron frequency $\omega_{Be} = \omega_{Be}^0(1 + \delta B/B_0)$.

The nonlinear correction to the magnetic field, δB, is derived by a method similar to that described above for electrostatic oscillations. When $E_\perp >> E_z$, $k_z >> k_\perp$, $k_\perp r_{Bj} < 1$, and $\omega \simeq n\omega_{Bj}$, for steady-state oscillations we have:

$$(1 - F_n)^2 F_n = A_n,$$

$$A_n = -\frac{n}{2(n-1)!}\frac{(\omega_{pj}^0)^2 n\,\omega_{Bj}^0}{(\omega - n\,\omega_{Bj}^0)^3}\frac{|P_{n-1,j}E_\perp|^2}{B_0^2}, \tag{4.29}$$

where $F_n = [n\,\omega_{Bj}^0/(\omega - n\,\omega_{Bj}^0)](\delta B/B_0)$. It follows from this that $\delta B < 0$, that is, the HF pressure produces a magnetic well.

These oscillations affect the slow nonlinear variations in the density most strongly through the Lorentz force, which makes the plasma move along the magnetic field. If the typical scale length L of the inhomogeneity in the amplitude of the electric field of the oscillations along the magnetic field is less than $v_{Tj}\tau$, where τ is the time during which the source emits the wave beam and v_{Tj} is the thermal speed of particles of species j along the magnetic field, then a static (slowly varying) regime is realized for both the electrons and ions. In that case, δn for steady-state oscillations near the n-th harmonic of the cyclotron frequency is equal to δn_s, where

$$\delta n_s = -\frac{n}{(n-1)!}\,\frac{\omega_{pj}^2}{\omega(\omega - n\omega_{Bj})}\,\frac{|P_{n-1}\;E_\perp|^2}{8\,\pi(T_{\parallel e} + T_{\parallel i})}\,. \tag{4.30}$$

Here the subscript j denotes the species of particles acted on by the nonlinear Lorentz force. Note that $\delta n_e = \delta n_i = \delta n$; that is the density varies in a quasi-neutral fashion.

If the wave pulse is sufficiently short, i.e. $\tau < L/v_{Tj}$, then the electrons and ions do not have time to acquire a Boltzmann distribution along the magnetic field within the wave beam. In this transient regime, when the quasistatic field of charge separation is taken into account, we obtain the following expressions for the averaged nonlinear variations of the electron and ion density:

$$\delta n_{de} = (\omega_{pi}^{-2}\partial_t^2 + 1)\frac{m_e}{m_i}\langle n_{de}^{(2)}\rangle\,,$$

$$\delta n_{di} = \frac{m_e}{m_i}\langle n_{de}^{(2)}\rangle\,, \tag{4.31}$$

where $\langle n_{de}^{(2)}\rangle$ satisfies the equation

$$\partial_t^2\langle n_{de}^{(2)}\rangle = -\frac{(T_{\parallel e} + T_{\parallel i})}{m_e}\,\partial_z^2\delta n_s. \tag{4.32}$$

Let us compare the effect of the HF diamagnetism and the nonlinear variation in the density for $\omega \simeq \omega_{Be}$. The quantity $\delta n_s/n_0$ is $\omega(\omega_{Be}^0 - \omega)(k_z v_{Te})^{-2} \gg 1$ times greater than F in the equation (4.28) for the electron cyclotron oscillations near the first harmonic ($n = 1$, $F = F_1$) (see (4.29) and (4.30)). For this reason, when the wave pulses are long enough $\tau > L/v_{Tj}$, so that a Boltzmann distribution can be established during the slow nonlinear variations of electron and ion density, the diamagnetic effect is not significant. Note that the departure from the steady state in (4.28) can be disregarded for $\tau > L/v_g$, where

v_g is the wave group velocity along the magnetic field. That condition is certainly satisfied in our case, since $v_g \gg v_{Tj}$.

If the wave pulse is short enough $\tau < L/v_{Tj}$, then it follows from (4.31) and (4.32) that $\delta n_{de} \sim \max\{(\omega_{pi}\tau)^{-2}, 1\}\,(c_s\tau/L)^2\delta n_s$, where c_s is the ion sound speed. Assuming that the pulse is moderately short $\omega_{pi}\tau > 1$, we find that the diamagnetic effect exceeds the nonlinear effect on the density in (4.28) if the following condition is satisfied:

$$\frac{(\omega - \omega_{Be}^0)^2}{(\omega_{pe}^0)^2}\,\frac{c^2\tau^2}{L^2}\,\frac{m_e}{m_i} < 1. \tag{4.33}$$

This condition, together with the inequality $\tau < L/v_{Tj}$, imposes a limitation upon the lifetime of a wave beam. On the other hand, the beam must be sufficiently long, $\tau > L/v_g$, for its behaviour to be described by the steady-state equation (4.28). It is readily verified that this inequality is compatible with (4.33). Note that in the intermediate case ($L/v_{Te} < \tau < L/v_{Ti}$) of the supersonic regime, one should substitute the transverse beam dimension R in place of L in the inequality (4.33).

Equation (4.28) together with (4.29) and (4.30), or equations (4.31) and (4.32) for first harmonic electron cyclotron oscillations, describe the self-focusing of a wave beam when the amplitude is large enough. Nekrasov and Petviashvili (1981) have shown that when diamagnetism is the principal nonlinear effect, self-focusing can occur for $\omega_{Be}^0 > \omega_{pe}^0$ ($\sigma > 0$). The typical longitudinal scale length for self-focusing, when the wave amplitude exceeds the threshold value E_{th}, is equal to (to within an order of magnitude)

$$L \sim (\omega - \omega_{Be}^0)^2(\omega_{pe}^0)^{-2}\,\frac{B_0}{|E_\perp|}\,R,$$

for $a \gtrsim 1$, where R is the beam radius. This length may be small compared to the typical scale length of the variations in the magnetic field. Thus, when a powerful wave produces cyclotron heating in an inhomogeneous magnetic field, the region at which heating takes place according to the linear theory may be shifted by a considerable distance because of self-focusing.

As equation (4.28) includes the nonlinearity in an unusual fashion, it was solved numerically for a wave beam with axial symmetry, taking account of the diamagnetic effect only when the condition (4.33) is satisfied (Nekrasov and Petviashvili, 1981). Figure 4.7 shows the evolution of the distribution of the electric field amplitude along the radius (self-focusing) when the initial beam radius is $R = 50\ q^{-1}$ and the excess above the threshold is small.

The diamagnetic effect and the nonlinear variation in the density can lead to 3D localization of a wave packet of plasma oscillations with the formation of a soliton envelope.

We now determine the dominant packet frequency ω_0 and the corresponding wave vector $\mathbf{k}_{0z}$. We assume the amplitude of the electric field varies little during the time taken by the packet to travel a distance equal to its characteristic dimension L along the magnetic field. A solution of the equation for the amplitude of the electric field, $\mathbf{E}_\perp$, can then be sought in the form $E_\perp = E_\perp(z - v_g t, \mathbf{r}_\perp, t)$. Taking account of the diamagnetic effect only when condition (4.33) is satisfied, where $\tau = \tau_0 = L/v_g$, in accordance with (4.27) we obtain

$$i\partial_\tau E_\perp + \nabla^2_{\vec{\rho}} E_\perp + \partial^2_\zeta E_\perp + \beta |E_\perp|^2 E_\perp = 0, \qquad (4.34)$$

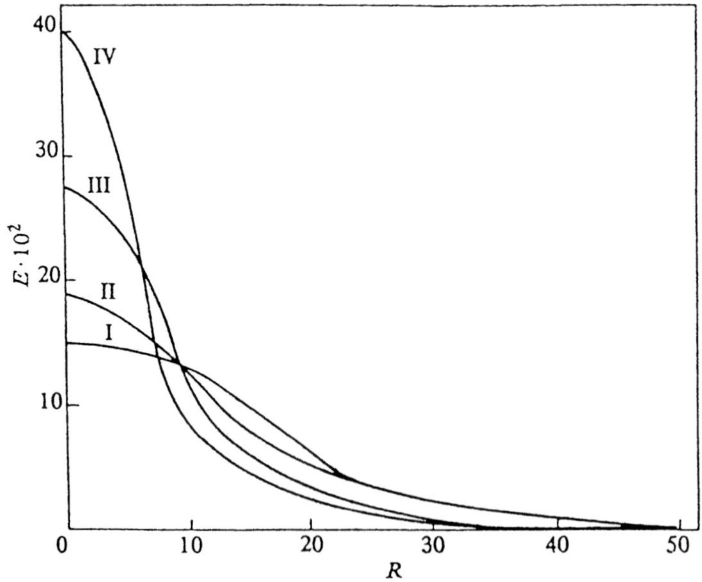

Figure 4.7. Evolution of the dimensionless amplitude distribution $E = \sqrt{A_1(E_\perp)}$ at a beam front along the radius R (R is expressed in q^{-1} units) as a function of z for $\alpha = 1$. The initial beam radius is $R_0 = 50\ q^{-1}$, the initial amplitude at the beam centre is $E_0 = 0.15$ ($E_c = 0.1$). Curve 1 corresponds to $z = 0$, II to $z = R_0$, III to $z = 2R_0$, IV to $z = (5/2)R_0$.

where

$$\tau = \omega_{Be}^0 t, \ \rho = \frac{\Omega}{c}(1+b)^{-1/2} r_\perp, \ \zeta = \frac{\Omega z}{c},$$

$$\Omega^2 = \frac{(\omega_{pe}^0 \omega_{Be}^0)^2}{(\omega_0 - \omega_{Be}^0)^2} + 2\omega_0 \omega_{Be}^0, \ b = -\frac{(\omega_{pe}^0)^2 \omega_{Be}^0}{2(\omega_0^2 - (\omega_{pe}^0)^2)(\omega_0 - \omega_{Be}^0)},$$

$$\beta = \frac{(\omega_{pe}^0)^4 \omega_0 \omega_{Be}^0}{2\Omega^2(\omega_0 - \omega_{Be}^0)^4 B_0^2}, \ v_g = \frac{2\,\omega_{Be}^0 k_{0z} c^2}{\Omega^2}.$$

Equation (4.34) is derived under the conditions $|\omega_0 - \omega_{Be}^0| \gg \partial_t$, $k_{0z} \gg \partial_z$, and $F \ll 1$. This equation is not a steady-state equation, in contrast to (4.28).

For steady-state solutions of the form $E_\perp \propto \exp(i\Gamma^2 \tau)$, equation (4.34) takes the form

$$\nabla^2 f = f - f^3. \tag{4.35}$$

Here $f = \beta^{1/2}\Gamma^{-1}|E_\perp|$, $\nabla^2 = \nabla^2_{\rho'} + \nabla^2_{\zeta'}$, $\rho' = \Gamma\rho$, $\zeta' = \Gamma\zeta$, and Γ is the dimensionless amplitude which obeys the relation $\omega_{Be}^0\Gamma^2 < v_g/L < |\omega_0 - \omega_{Be}^0|$ under our assumptions.

Equation (4.35) has soliton solutions (in particular, spherically symmetric ones) with typical dimensions of order unity. If the frequency of the oscillations, ω_0, satisfies $\omega_0|\omega_0 - \omega_{Be}^0| \lesssim (\omega_{pe}^0)^2$, then their amplitude is $E_{\perp 0} \sim \beta^{-1/2}\Gamma < |\omega_0 - \omega_{Be}^0|^{3/2} B_0/\omega_{pe}^0(\omega_{Be}^0)^{1/2}$. To within an order of magnitude the characteristic dimensions of the envelope soliton are

$$R \sim \frac{|\omega_0 - \omega_{Be}^0|^{3/2}}{(\omega_{Be}^0)^{3/2}} \frac{c}{\omega_{pe}^0} \frac{B_0}{E_{\perp 0}},$$

$$L \sim \frac{(\omega_0 - \omega_{Be}^0)^2}{(\omega_{pe}^0)^2} \frac{c}{\omega_{Be}^0} \frac{B_0}{E_{\perp 0}}.$$

Note that in this case the condition (4.33) has the form $(m_e/4m_i) \times [\omega_{Be}^0/|\omega_0 - \omega_{Be}^0|] < 1$. This is easily satisfied. Thus, diamagnetism is the principal nonlinear effect that produces an envelope soliton for the 3D cyclotron wave packets under discussion here.

We should note, however, that the soliton solution of (4.34) is unstable by the Vakhitov–Kolokolov criterion (3.29). If the energy carried by a wave packet exceeds that of a similar-sized soliton, then the packet will collapse, while if the energy is smaller, the packet will spread.

4.5. DIAMAGNETIC SELF-FOCUSING OF ELECTROMAGNETIC CYCLOTRON WAVES TRAVELLING ACROSS A MAGNETIC FIELD

Electromagnetic waves that travel nearly perpendicular to the ambient magnetic field are frequently used in radio frequency heating of plasmas. In that case, two different electromagnetic waves exist in the electron cyclotron frequency range in the plasma—the extraordinary and ordinary waves. The second harmonic of the electron cyclotron frequency is usually employed for heating by the extraordinary wave, while the first harmonic is used for heating by the ordinary wave. Obviously, a study of the influence of nonlinear effects on the propagation of these waves is of practical interest.

Nekrasov (1986) has considered self-focusing of extraordinary and ordinary cyclotron waves associated with the diamagnetic effect. The nonlinear variation in the density was taken into account and the conditions under which it can be disregarded were given.

First, we discuss self-focusing of the extraordinary wave. When $k_\perp \gg k_z$, we have $E_\perp \gg E_z$, where $E_z \sim k_z E_\perp / k_\perp$. Here E and k are the electric field and the wave vector of oscillations. The z-axis is directed along the magnetic field $\mathbf{B}_0$. Assuming that $\omega - n\omega_{Bj} \gg k_z v_{T\|j}$, $n\,\omega_{Bj}(v_{Tj}/c)^2$ and $k_\perp r_{Bj} \ll 1$, where v_{Tj} is the thermal velocity of the particles, we find the nonlinear variation of the magnetic field δB to be

$$\delta B = -\sum_{j,\, n>0} \frac{n^2 \omega_{pj}^2}{2n!B_0} \langle |\chi_+^{(n)}|^2 \rangle, \tag{4.36}$$

where

$$\chi_+^{(n)} = P_{n-1,j} \sum_{\omega} \frac{E_\omega^+}{\omega - n\,\omega_{Bj}} \exp(-i\,\omega t).$$

Here $E^+_\omega = E_{x\omega} + iE_{y\omega}$ is the Fourier component of the electric field.

From (4.36) one can see that the magnetic field decreases in the localization region of a wave packet. Note that (4.36) is valid for any frequency.

Consider the self-focusing associated with the diamagnetic effect alone. For an extraordinary electromagnetic ($N_\perp \simeq 1$) wave this is possible either at the initial stage of self-focusing or when the wave pulse is short enough, so that the nonlinear perturbation in the plasma density does not have enough time to develop a Boltzmann distribution. Let τ be the time during which the electromagnetic field of the wave acts along a given line of force of the ambient magnetic field during diamagnetic self-focusing. We shall assume that

$$\frac{L_z}{v_{T\|e}} \ll \tau \ll \frac{L_z}{v_{T\|i}}; \quad \tau \ll \frac{L_z}{c_s}, \tag{4.37}$$

where L_z is the typical size of the wave beam along the magnetic field, c_s is the ion sound speed. The second of conditions (4.37) corresponds to a supersonic regime. Note that the focusing of an electromagnetic wave with frequency $\omega \sim k_\perp c$ occurs during a time $\tau_f \sim L_f / c$, where L_f is the focusing length.

We now find the nonlinear variation in the density. We shall assume that the high-frequency oscillations involve the electrons. Calculating the nonlinear variation in the density, δn, with the nonlinear quasi-static potential taken into account, we obtain the following equations when the conditions (4.37) and $\omega_{Bi}\tau \gtrsim 1$ are satisfied:

$$\{\partial_t^2 \delta n = \frac{e^2 n_0}{4 m_e m_i} \partial_z^2 \sum_{n>0} \frac{n^2}{n!} \langle \chi_+^{(n)} \psi_+^{(n)*} + \chi_+^{(n)*} \psi_+^{(n)} \rangle , \tag{4.38}$$

when $\omega_{Bi}\tau > L_z / L_\perp$, and

$$\delta n = \frac{e^2 n_0}{4 m_e m_i \omega_{Bi}^2} \nabla_\perp^2 \sum_{n>0} \frac{n^2}{n!} \langle \chi_+^{(n)} \psi_+^{(n)*} + \chi_+^{(n)*} \psi_+^{(n)} \rangle \tag{4.39}$$

when $\omega_{Bi}\tau < L_z / L_\perp$, where

$$\psi_+^{(n)} = P_{n-1,e} \cdot \sum_\omega \frac{E^+_\omega}{\omega} \exp(-i\omega t).$$

In these equations $n_0 = n_{0j}$ is the unperturbed density, $\delta n = \delta n_i =$

δn_e, $L_\perp$ is the typical transverse dimension of the wave beam perpendicular to the magnetic field. The variation δn in the density under the above conditions is associated with the establishment of a Boltzmann distribution for the electrons in the quasi-static potential field and in the high-frequency potential field. Ions in a quasi-static potential move along the magnetic field in the case (4.38) and across the magnetic field in the case (4.39). When $N_\perp \sim 1$, the principal contribution to (4.38) and (4.39) is from the first harmonic ($n = 1$), and that to (4.36) is from the first and second harmonics ($n = 1,2$), with the second harmonic providing a greater contribution when $(\omega - 2\,\omega_{Be})/\omega_{Be} < v_{T\perp e}/c$, where $v_{T\perp e}$ is the transverse electron thermal speed.

We now derive a nonlinear wave equation for an extraordinary electromagnetic wave, for which $N_\perp \sim 1$. We represent the electric field of the wave in the form

$$\mathbf{E} = \frac{1}{2}\mathbf{E}_0 \exp(-i\,\omega_0 t + ik_{0x}x) + c.c.$$

Here k_{0x} and ω_0 are the dominant wave vector and frequency in the packet and are related by

$$k_{0x}^2 c^2 = \omega_0^2 - (\omega_{pe}^0)^2 + \frac{(\omega_{pe}^0\omega_{Be}^0)^2}{(\omega_{pe}^0)^2 + (\omega_{Be}^0)^2 - \omega_0^2}\,.$$

We consider the case in which the wave frequency is close to the second harmonic of the electron cyclotron frequency. The magnetic field B_0 is assumed to be inhomogeneous in the transverse direction with the cyclotron resonance occurring at $x = 0$ in the linear approximation. We thus have

$$2\,\omega_{Be}(x) = \omega_0(1 + x/L + h),$$

where the relative nonlinear change in the magnetic field, h, has been included.

Consider the region where

$$|x/L + h| > \max\left\{\frac{v_{T\perp e}}{c},\ \left(\frac{\omega_{pe}^0}{\omega_{Be}^0}\,\frac{v_{T\perp e}v_{T\|e}}{c^2}\right)^{2/3}\right\}. \tag{4.40}$$

In this case the main contribution to h is from the first harmonic ($n = 1$). Then, as shown by Nekrasov (1986), the nonlinear variation in the density can be disregarded in the wave equation under the conditions

$$\frac{\omega_{pe}^0}{\omega_{Be}^0} > \left(\frac{m_e}{m_i}\right)^{1/2} \frac{\tau c}{L_z}, \quad \omega_{Bi}^0 \tau > \frac{L_z}{L_\perp},$$

$$\frac{\omega_{pe}^0}{\omega_{Be}^0} > \left(\frac{m_e}{m_i}\right)^{1/2} \frac{c}{\omega_{Bi}^0 L_\perp}, \quad \omega_{Bi}^0 \tau < \frac{L_z}{L_\perp}.$$

The nonlinear equation for the amplitude $E_0^- \equiv E_{x0} - iE_{y0}$ then becomes

$$2ik_{0x}\partial_x E_0^- + \partial_y^2 E_0^- + \mu_1 \partial_z^2 E_0^- = -(\mu_2/B_0^2)|E_0^-|^2 E_0^-, \tag{4.41}$$

where

$$\mu_1 = \frac{\omega_0^2(\omega_0^2 - (\omega_{Be}^0)^2) - (\omega_{pe}^0)^4}{(\omega_0^2 - (\omega_{Be}^0)^2 - (\omega_{pe}^0)^2)}, \quad \mu_2 = \frac{(\omega_{pe}^0)^4(4(\omega_{Be}^0)^2 - (\omega_{pe}^0)^2)}{4(3(\omega_{Be}^0)^2 - (\omega_{pe}^0)^2)^2 c^2}.$$

An estimate of the self-focusing length L_f from (4.41) for $\omega_{pe}^0 \simeq \omega_{Be}^0$ gives ($\mu_1, \mu_2 > 0$):

$$L_f \sim 10(B_0/E_0)L_\perp. \tag{4.42}$$

It follows from the dispersion relation that the extraordinary electromagnetic wave may transform into the electrostatic Bernstein mode, for which $N_\perp \gg 1$, near the point of cyclotron resonance (depending on the direction of propagation for fixed ω_{pe}^0 and ω_{Be}^0). Nekrasov (1986) had examined the diamagnetic focusing of an electrostatic wave at the second harmonic of the electron cyclotron frequency. The nonlinear variation in the density is found for the case $\tau > L_z/v_{T\|j}$ in which the particles of type j have time to acquire a Boltzmann distribution. It was shown that the nonlinear variation in the density can be neglected in the wave equation for the potential amplitude in the region

$$\frac{(\omega_{pe}^0)^2}{4(\omega_{Be}^0)^2}\frac{v_{T\perp e}^2}{c^2} < \frac{|x|}{L} < \frac{\omega_{pe}^0}{\omega_{Be}^0}\frac{v_{T\parallel e}}{c}, \quad \text{when} \quad \left(\frac{\omega_{pe}^0}{\omega_{Be}^0}\right)^2 < 3,$$

$$\frac{v_{T\perp e}^2}{c^2} < \frac{x}{L} < \left(\frac{\omega_{pe}^0}{\omega_{Be}^0}\right)^2 \frac{v_{T\parallel e}}{c}, \quad \text{when} \quad \left(\frac{\omega_{pe}^0}{\omega_{Be}^0}\right)^2 > 3. \tag{4.43}$$

The left-hand inequalities in (4.43) define the region where electrostatic oscillations exist ($x < 0$ for $(\omega_{pe}^0/\omega_{Be}^0)^2 < 3$ and $x > 0$ for $(\omega_{pe}^0/\omega_{Be}^0)^2 > 3$).

An estimate of the diamagnetic self-focusing length L_f for $\omega_{pe}^0 \sim \omega_{Be}^0$ gives

$$L_f \sim L_z \left(\frac{T_{\perp e}}{T_{\parallel e}}\right)^{1/2} \left(\frac{|x_f|}{L}\right)^{3/2} \frac{B_0}{E_0}, \tag{4.44}$$

where x_f is the place at which focusing starts and it is assumed that $|x_f| > L_f$. Note that the self-focusing length (4.44) is considerably shorter than in (4.42).

Besides waves that are polarized perpendicular to the magnetic field, Nekrasov (1986) has considered the diamagnetic self-focusing of an ordinary wave polarized along the magnetic field and propagating transversely. When $k_\perp \gg k_z$, we have $E_z \gg E_\perp$, where $E_\perp \sim k_z E_z / k_\perp$. Let us assume that temperature anisotropy is not excessively large, i.e. $T_{\parallel j}/T_{\perp j} \gg |\omega/n\,\omega_{Bj} - 1|$. Under the same conditions as for the extraordinary wave, we obtain a nonlinear variation in the magnetic field near the first harmonic of the electron cyclotron frequency:

$$\delta B = -\frac{(\omega_{pe}^0)^2}{B_0}\frac{T_{\parallel e}}{T_{\perp e}}\langle|\chi_z^{(1)}|^2\rangle, \tag{4.45}$$

where in the general case

$$\chi_z^{(n)} = P_{nj}\sum_\omega \frac{E_{z\omega}}{\omega - n\,\omega_{Bj}} \exp(-i\,\omega t).$$

Formula (4.45) is true under the condition

$$\frac{T_{\parallel e}}{T_{\perp e}} \frac{\omega}{|\omega - \omega_{Be}|} k_\perp r_{Be} \gg 1,$$

when the contribution of the zeroth harmonic ($n=0$) can be disregarded. Note that the appearance of the magnetic well (4.45) is in this case associated with the wave magnetic field.

The nonlinear variation in the density δn is determined from the equations ($\tau > L_z/v_{T\|e}$, $\omega_{Bi}^0 \tau \gtrsim 1$):

$$\partial_t^2 \delta n = \frac{e^2 n_0}{2m_e m_i (\omega_{Be}^0)^2} \partial_z^2 \langle |E_z|^2 \rangle, \text{ when } \omega_{Bi}^0 \tau > L_z/L_\perp \tag{4.46}$$

and

$$\delta n = \frac{e^2 n_0}{2m_e m_i (\omega_{Be}^0 \omega_{Bi})^2} \nabla_\perp^2 \langle |E_z|^2 \rangle, \text{ when } \omega_{Bi}^0 \tau < L_z/L_\perp. \tag{4.47}$$

Suppose the ambient magnetic field is a function of x, and the wave frequency ω_0 is identical with the electron cyclotron frequency at $x=0$ in the linear approximation:

$$\omega_{Be}(x) = \omega_0(1 + x/L + h).$$

We set

$$E_z = \frac{1}{2} E_{z\,0} \exp(-i\,\omega_0 t + ik_{0x}\,x) + c.c.$$

Then we get a nonlinear wave equation for the amplitude $E_{z\,0}$ in the region $|x|/L > |h|$ that includes the diamagnetic effect:

$$2ik_{0\,x}\partial_x E_{z\,0} + \partial_y^2 E_{z\,0} + \mu_3 \partial_z^2 E_{z\,0} = -(\mu_4/B_0^2)|E_{z\,0}|^2 E_{z\,0}, \tag{4.48}$$

where k_{0z} is given by $k_{0x}^2\, c^2 = \omega_0^2 - (\omega_{pe}^0)^2$. Equation (4.48) contains the following notation:

$$\mu_3 = \left[1 - \left(\frac{\omega_{pe}^0}{\omega_{Be}^0}\right)^2 \left(\frac{L}{x}\right)^3 \frac{v_{T\|e}^4}{c^4}\right] \frac{k_{0x}^2 c^2}{\omega_0^2} \quad \text{and} \quad \mu_4 = \left(\frac{\omega_{pe}^0}{\omega_{Be}^0}\right)^4 \frac{k_{0x}^4 v_{t\|e}^4}{16c^2\omega_0^2} \left(\frac{L}{x}\right)^4 .$$

The nonlinear variation in the density is not included in (4.48). This is possible under the conditions

$$\frac{\omega_{pe}^0}{\omega_{Be}^0} \frac{k_{0x}^2 v_{T\|e}^2}{(\omega_{Be}^0)^2} \left(\frac{L}{x}\right)^2 > \left(\frac{m_e}{m_i}\right)^{1/2} \frac{\tau c}{L_z}, \quad \text{when} \quad \omega_{Bi}^0 \tau > \frac{L_z}{L_\perp},$$

or

$$\frac{\omega_{pe}^0}{\omega_{Be}^0} \frac{k_{0x}^2 v_{T\|e}^2}{(\omega_{Be}^0)^2} \left(\frac{L}{x}\right)^2 > \left(\frac{m_e}{m_i}\right)^{1/2} \frac{c}{\omega_{Bi}^0 L_\perp}, \quad \text{when} \quad \omega_{Bi} \tau < \frac{L_z}{L_\perp}.$$

We now estimate the self-focusing length L_f associated with the inhomogeneity of the field amplitude along the y-axis. For $\omega_{Pe}^0 \simeq \omega_{Be}^0$, we have

$$L_f \sim 8L_\perp \left(\frac{x_f}{L}\right)^2 \frac{c^2}{v_{T\|e}^2} \frac{B_0}{|E_{z\,0}|},$$

where $|x_f| > L_f$.

Note that, if $\tau > L_z/v_{T\|j}$ (both for electrons and ions), then δn can be neglected in (4.48) in the region

$$\frac{|x|}{L} < \left(\frac{\omega_{pe}^0}{\omega_{Be}^0}\right)^{1/2} \frac{k_{0x} c}{\omega_0} \left(\frac{v_{T\|e}}{c}\right)^{3/2} .$$

Therefore, the diamagnetic effect is stronger than the maximum density perturbation in this region.

4.6. EFFECTS OF CYCLOTRON SOLITONS

Electrostatic cyclotron waves can be excited in plasmas for various reasons, for example, by beams of particles along the magnetic field, as well as in the presence of fast particles, moving across the magnetic field when the distribution function contains a loss cone.

We briefly discuss the possible kinetic effects in plasmas owing to the appearance of cyclotron solitons in cases where the plasma pressure is lower than the pressure of the ambient magnetic field. Since the HF pressure in a soliton is balanced by the magnetic pressure, rather than by the plasma pressure, the former may approach or even exceed the plasma pressure. Expulsion of plasma from the soliton may be limited because of its great length along the magnetic field, as well as because of a longitudinal inhomogeneity in the magnetic field (owing to formation of magnetic mirrors). Because cyclotron waves are capable of self-localization in the shape of solitons, the wave energy density can become large. This may have been observed in experiments on RF plasma heating as a broadening of spectral lines emitted by the plasma.The high energy density of the electric oscillations in a soliton may lead to intensive heating and anomalous resistivity. In such cases the energy of the wave electric field is converted into transverse kinetic energy of trapped electrons or ions owing to the cyclotron resonance. This mechanism may have been a source of particles with high transverse energies recorded in discharges within runaway electrons (TFR Group, 1975).

The formation of cyclotron solitons may explain a highly nonuniform spatial distribution of cyclotron oscillations observed by Mozer *et al.* (1977) in the auroral region. The jump in the distribution is so sharp that one gets the impression of a shock wave with its front along $\mathbf{B}_0$. The relative amplitude of the density oscillations reaches 20 to 30 per cent.

Electromagnetic cyclotron waves are used for plasma heating in the laboratory (TFR Group, 1975).They have also been used extensively for heating the ionospheric plasma.

The main nonlinear effect for short wave pulses is HF diamagnetism. If the pulses are long enough, the nonlinearity is usually caused by a quasi-neutral density variation, when the particles have time to acquire a Boltzmann distribution. However, for the Bernstein mode ($N_\perp \gg 1$) and an ordinary wave travelling across the magnetic field, it is possible for diamagnetism to be the main nonlinear effect even when the density obeys a Boltzmann distribution. The self-focusing length of the Bernstein mode turns out to be considerably shorter than the self-focusing lengths of electromagnetic waves with refractive indices on the order of unity. Because of this, an extraordinary wave travelling across a magnetic field can be focused and absorbed by the plasma after conversion into the Bernstein mode.

5. Solitary vortices in the atmospheres of rapidly rotating planets

5.1. STRUCTURES IN ATMOSPHERES AND OCEANS

Nonuniform heating in the planetary atmospheres produces winds with different structures. Weather is largely controlled by large-scale winds. It has been noticed that they mostly blow latitudinally in the rapidly rotating planets (in the zonal direction). Examples of such winds are trade winds and monsoons in the tropical and subtropical latitudes. Zonal flows can be discerned especially clearly in the atmosphere of Jupiter, where they change direction several times from the equator to the pole (see Figure 5.1) (Smith *et al.*, 1979 a,b). Zonal propagation of flows in the oceans is prevented by the presence of continents. The roughness of the Earth's surface also makes the flows deviate from a zonal pattern.

The planetary rotation gives rise to the Coriolis force which acts on the moving masses and could make them rotate in the opposite sense from planetary rotation (anticyclonically). However, a detailed consideration of this problem shows that rotation stabilizes zonal flows, even when they are inhomogeneous. Large-scale (synoptic) vortices, cyclones or anticyclones, are observed against a background of zonal flows. A cyclone is a vortex which rotates in the same sense as the planet's rotation, while air in an anticyclone circulates in the opposite sense. Large vortices are drifting at a velocity less than or of the order of the rotation velocity in the vortex. The drift mostly occurs in the opposite sense from the planet's rotation. The vorticity is smaller than the angular rotation velocity of the planet ω_0. However, topical cyclones whose vorticity approaches ω_0 sometimes appear in subtropical zones. In the particularly strong vortices known as tornadoes, the vorticity is large compared to ω_0. However, their

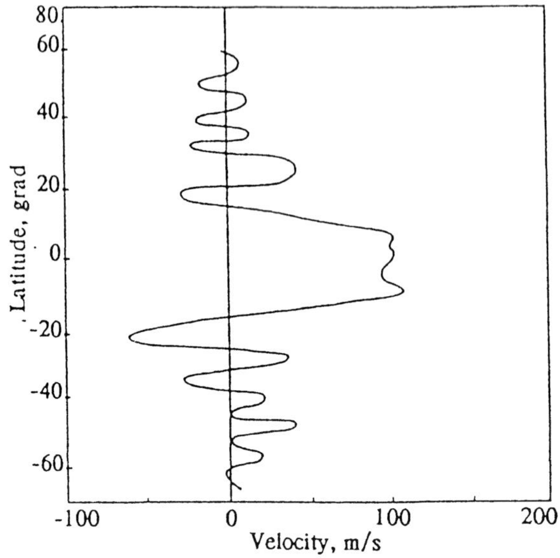

Figure 5.1. The distribution of the zonal flow velocity over the latitudes of Jupiter. The high velocity in the equatorial region is probably due to a differential rotation of the planet.

dimensions are typically small compared to the height of the atmosphere. In the following discussion we shall only consider vortices whose dimensions are large compared to the height of the atmosphere, because they can be described in the so-called shallow-atmosphere approximation. In these vortices the pressure obeys the hydrostatic approximation, while the vertical velocity is small compared to the horizontal velocity.

5.2. EQUATIONS FOR SHALLOW WATER

In a number of cases, large-scale flows in the atmosphere can be adequately described by the equations for shallow water. They are derived from the 3D equations of hydrodynamics assuming a constant density and a small depth compared to the horizontal scale length of the perturbations

$$d_t \mathbf{v} = -\nabla p/\rho - g\zeta, \quad d_t \equiv \partial_t + \mathbf{v}\cdot\nabla, \tag{5.1}$$

$$\operatorname{div} \mathbf{v} = 0. \tag{5.2}$$

According to (5.2), the vertical component of the velocity in a shallow fluid is small compared to the horizontal component. For this reason, the left-hand side in the vertical component of (5.1) can be neglected. We then get

$$\partial_z p = -g\rho, \quad p = -\rho g\,(z - H_1) + p_0, \tag{5.3}$$

where H_1 is the deviation of the fluid depth from equilibrium, P_0 = constant is the pressure at the fluid surface, and the z coordinate is measured from the undisturbed surface. The horizontal component of (5.1) then takes the form

$$d_t \mathbf{v}_\perp = -g\nabla_\perp H_1. \tag{5.4}$$

Shallow water is equivalent to a 2D compressible gas with a depth of $H = H_1 + H_0$, where H_0 is the depth of the undisturbed fluid, which may be a function of the coordinates. The equation of continuity for such a gas is

$$\partial_t H + \operatorname{div} \mathbf{v}_\perp H = 0, \quad H = H_1 + H_0. \tag{5.5}$$

The effect of roughness of the bottom of the fluid is incorporated in the undisturbed depth $H_0(x, y)$. The system of equations (5.4), (5.5) is referred to as the shallow water equations.

When the fluid rotates, it is convenient to transform to a rotating reference frame. This is equivalent to including the Coriolis force in the equation of motion (5.4):

$$d_t \mathbf{v}_\perp = -g\nabla_\perp H_1 + \Omega\,(\mathbf{v}_\perp \times \vec{\zeta})\,, \tag{5.6}$$

where Ω is the Coriolis parameter. This does not affect equation (5.3), because the vertical component of the Coriolis force can be neglected compared to gravity. It is generally accepted that the set (5.5), (5.6) adequately describes large-scale disturbances in the atmosphere and ocean. However, as will be apparent from the next section, including the compressibility makes a significant contribution to the dynamics of the atmosphere. Hence the equations for a shallow atmosphere are somewhat different. The system of (5.5), (5.6) has a large number of first integrals. From these one can, in particular, derive an equation for the frozen generalized vorticity,

$$d_t q = 0, \quad q = (\zeta \cdot \operatorname{curl} \mathbf{v} + \Omega)/H \tag{5.7}$$

The following integrals are also conserved:

$$E = \frac{1}{2}\int H(v^2 + gH)d^2x, \tag{5.8}$$

$$M_z = \int \{(\mathbf{r} \times \mathbf{v}_\perp)_z + \Omega\} H d^2x, \tag{5.9}$$

$$I = \int H\Phi(q)d^2x, \tag{5.10}$$

where the first two integrals are the total energy and the projection of the angular momentum along the meridian. The latitudinal component of M is not conserved owing to the inhomogeneity of Ω (it depends on the latitude). In (5.10) $\Phi(q)$ is an arbitrary function and the integral is taken over an arbitrary region D whose points move at a velocity $\mathbf{v}_\perp$ (over the fluid region).

The above equations are invariant under changes in the shape of the surface: they can be used to describe waves on a plane or on the surface of a sphere.

In the preceding equations it was assumed that horizontal velocity does not depend on z. This corresponds to perturbations averaged over a short scale length and over short times on the order of or less than the period of rotation of the planet. Eckman pointed out that the Coriolis force smooths out the dependence of the velocity on height. This can be shown by a simple example: a constant pressure gradient $\mathbf{A} = -\nabla p/\rho = \text{const}$ exists in the atmosphere. Then, assuming $v_z = 0$ and that the horizontal velocity depends only on z, near the planet's surface, we have

$$\mathbf{A} + \Omega(\zeta \times \mathbf{v}) + \nu\partial_z^2\mathbf{v} = 0,$$

where the boundary condition is $\mathbf{v} = 0$ at $z = 0$ and ν is the kinematic viscosity coefficient. In the absence of a Coriolis force ($\Omega = 0$), the velocity decreases with altitude as z^2. If $\Omega \neq 0$, this equation has a solution with a localized boundary layer of thickness $l = (2\nu/\Omega)^{1/2}$:

$$v_x = (A/\Omega)\exp(-z/l)\sin(z/l),$$

$$v_y = (A/\Omega)\exp(-z/l)\cos(z/l) - 1.$$

Outside the Eckman boundary layer the velocity becomes constant and perpendicular to A. For the Earth's atmosphere $l \simeq 60$ cm (for the ocean, 5 cm). The shallow water approximation is, therefore, valid outside the tropical zone for atmospheric perturbations with vertical dimensions much longer than 50 cm.

5.3. EQUATIONS FOR A SHALLOW ATMOSPHERE

The equations for waves on shallow water have been long known and studied. It is enough to note that the Korteweg–de Vries equation was derived from them. It would seem that large-scale disturbances in the atmosphere ought to obey similar equations. As we shall see,however, there are certain peculiarities in this case. Long waves in the atmosphere are commonly considered to obey the shallow water equations (5.5)–(5.6), with the pressure and temperature variations assumed to be similar. As one can see from synoptic maps, however, that assumption is not always true. We give a simple derivation of shallow atmosphere equations which includes that difference (Petviashvili and Pokhotelov, 1988). In a rotating reference frame attached to the planet, the equations for the atmosphere are

$$d_t \mathbf{v} = -\nabla p/\rho - g\zeta + \Omega(\mathbf{v}\times\zeta), \tag{5.11}$$

$$d_t \rho + \rho \operatorname{div} \mathbf{v} = 0, \tag{5.12}$$

and

$$d_t p + \gamma p \operatorname{div} \mathbf{v} = 0, \tag{5.13}$$

where γ is the ratio of specific heats. For the sake of simplicity, the effects of heating and dissipative processes are neglected in the pressure equation (5.13). In the hydrostatic approximation the pressure of the atmosphere at the bottom is determined from the z-component of (5.11) as

$$p_b = g\int_0^\infty \rho dz, \tag{5.14}$$

where b corresponds to the value at the bottom of the atmosphere ($z = 0$).

The equation of continuity for the atmospheric density at the surface of the planet is, with $v_{zb} = 0$ taken into account,

$$\partial_t \rho_b + \mathbf{v}_\perp \cdot \nabla \rho_b + \rho_b \operatorname{div}|_{\zeta=0} \mathbf{v} = 0. \tag{5.15}$$

Further, integrating (5.12) using (5.14) we get

$$\partial_t p_b + \mathrm{div}|_{\zeta=0}\mathbf{v}_\perp p_b = 0. \tag{5.16}$$

Here, $\mathbf{v}_\perp$ is assumed to depend weakly on z (as averaged over time interval Ω^{-1}).

Equation (5.13) at the bottom gives:

$$\partial_t p_b + \mathbf{v}_\perp \cdot \nabla p_b + \gamma p_b \mathrm{div}|_{\zeta=0}\mathbf{v} = 0. \tag{5.17}$$

Comparing (5.17) and (5.16), we find div **v** at the surface of the planet to be

$$\mathrm{div}|_{\zeta=0}\mathbf{v} = \gamma^{-1}\mathrm{div}|_{\zeta=0}\mathbf{v}_\perp, \tag{5.18}$$

which, on substitution into (5.15), yields

$$\partial_t \rho_b^\gamma + \mathrm{div}|_{\zeta=0}\mathbf{v}_\perp \rho_b^\gamma = 0. \tag{5.19}$$

Finally, we write the horizontal component of the equation of motion (5.11) at $z = 0$:

$$\partial_t \mathbf{v}_\perp + \mathbf{v}_\perp \cdot \nabla \mathbf{v}_\perp = -\nabla_\perp p_b/\rho_b + \Omega(\mathbf{v}_\perp \times \zeta). \tag{5.20}$$

Equations (5.16), (5.19), (5.20) make a complete system of equations for a shallow atmosphere. One can see that the compressibility of the atmosphere cannot be neglected in long waves, because γ is finite. This system of equations transforms into the shallow water equations (5.5) and (5.6) when $\gamma \to \infty$ and $\rho_b = \text{const}$. Here we have $H = \int_0^\infty \rho dz/\rho_b = p_b/g\rho_b$.

5.4. THE GEOSTROPHIC APPROXIMATION

We now consider low-frequency oscillations whose frequencies are small compared to Ω. The equations for a shallow atmosphere can then be simplified. To do this, we use the so-called geostrophic approximation, i.e., we expand in powers of ω/Ω, where ω is the oscillation frequency. In this approximation we have, from (5.11),

$$\mathbf{v}_\perp = \mathbf{v}_g + \mathbf{v}_i + \ldots, \quad \mathbf{v}_g = (\rho\Omega)^{-1}(\zeta\times\nabla p),$$
$$\mathbf{v}_i = \Omega^{-1}(\zeta\times(\partial_t + \mathbf{v}_g\cdot\nabla)\mathbf{v}_g). \tag{5.21}$$

The subscript b is omitted here and in the following. Substituting (5.21) into (5.16) and (5.19), we obtain (Petviashvili and Pokhotelov, 1988)

$$\partial_t(p - r_R^2\nabla^2 p) - \frac{r_R^4}{p_0}\{p, \nabla^2 p\} + v_R\partial_x p - \frac{r_R^2\Omega}{\rho_0}\{p, \rho\} = 0 \tag{5.22}$$

and $v_R = c_s^2\Omega^{-1}\partial_y \ln(\rho_0/\Omega)$; $r_R^2 = p_0/\rho_0\Omega^2$

$$\partial_t\rho + \frac{(\gamma - 1)\Omega r_R^2}{\gamma p_0}\{p, \rho\} + \frac{v_R}{c_s^2}\partial_x P = 0, \tag{5.23}$$

where $c_s = (\gamma p_0/\rho_0)^{1/2}$ is the speed of sound in the atmosphere $\{p,\rho\} \equiv \partial_x p\partial_y\rho - \partial_y p\partial_x\rho$, and $\{\ldots, \ldots\}$ denotes the Jacobian.

When expressed in dimensionless form, these equations become

$$\partial_t(p - \nabla^2 p) + u_R\partial_x p - \{p, \nabla^2 p\} = \{p, \rho\}, \tag{5.24}$$

and

$$\partial_t\rho + \frac{\gamma - 1}{\gamma}\{p, \rho\} + u_R\gamma^{-1}\partial_x p = 0. \tag{5.25}$$

Here, time is measured in units of Ω^{-1} and the coordinates in units of the Rossby radius $r_R = (p_0/\rho_0)^{1/2}/\Omega$, u_R is the nondimensionalized Rossby velocity $u_R = v_R/(p_0/\rho_0)^{1/2}$, and the pressure and density are in units of p_0 and ρ_0, respectively. (5.24) and (5.25) have been transformed approximately into cartesian coordinates with the x-axis pointing eastward and the y-axis pointing northward. When $\gamma \to \infty$ and $\rho = \text{const}$, this system of equations becomes

$$\partial_t(p - \nabla^2 p) + u_R\partial_x p = \{p, \nabla^2 p\}, \tag{5.26}$$

which was proposed by Charney (1947) to describe Rossby waves in the atmosphere. The set (5.22) and (5.23), as well as the Charney equation (5.26), can be applied to disturbances whose dimensions are on the order of or less than the Rossby radius. For large scales, as shown in § 5.8, one should include the variability of the coefficients. Hasegawa and Mima (1978) and Hasegawa (1985) have derived a corresponding equation for electrostatic drift waves in plasma.

The atmosphere density ρ is related to the pressure and temperature T by the equation of state for an ideal gas:

$$p = \rho T, \tag{5.27}$$

where all the quantities are dimensionless.

It follows from this that the Charney equation is applicable when the isotherms and isobars are sufficiently close to one another. Otherwise, low-frequency disturbances are more conveniently described by equations (5.24) and (5.25).

5.5. ZONAL FLOWS AND THEIR STABILITY

The simplest nontrivial solution to the equations for a shallow atmosphere is zonal flow (that is, a latitudinal flow owing to a nonuniform pressure along a meridian). This solution is independent of time and has the form

$$v_* = -(\rho\Omega)^{-1}\partial_y p, \quad v_y = 0, \tag{5.28}$$

for arbitrary functions $p = p(y)$ and $\rho = \rho(y)$.

Zonal flows are observed in Jupiter's atmosphere (Figure 5.1) and in the trade winds and monsoons of the Earth's atmosphere. The zonal flows on Jupiter clearly change direction several times from the equator to the pole and are rather inhomogeneous. In the absence of the Coriolis force, this inhomogeneity could not exist because of instability. For this reason, studies of the stability of zonal flows are of great interest.

The stability theory of zonal flows is based on Rayleigh's theorem. Rayleigh investigated the stability of a plane-parallel flow of incompressible fluid. In the linear approximation perturbations of such flows obey the Rayleigh equation (Lin, 1955)

$$\psi'' - k^2\psi - \frac{u''}{u - c}\psi = 0. \tag{5.29}$$

Here $u(y)$ is the velocity of the unperturbed flow along x. The y-component of the velocity perturbation is written in the form $v_y =$

$ik\psi(y)\exp[ik(x-ct)]$. To investigate the stability of the flow, we multiply (5.29) by ψ^* (*denotes complex conjugation) and integrate over the cross-section of the flow assuming that the perturbation vanishes at the boundary. From (5.29) we then obtain

$$\int \{|\psi'|^2 + k^2|\psi|^2 + u''|\psi|^2/(u-c)\}dy = 0 \tag{5.30}$$

where c is a complex-valued quantity in general. The imaginary part of (5.30) then gives

$$\mathrm{Im}\, c\int (u''|\psi|^2/|u-c|^2)dy = 0. \tag{5.31}$$

When an instability is present $\mathrm{Im}\, c \neq 0$ and the condition (5.31) can be satisfied if the integrand changes sign. This is possible only when the flow profile has an inflection point, that is, when $u'' = 0$ at least at a single point. One can state that the flow is stable, if there are no inflexion points (Rayleigh's theorem).

The stability of zonal flows can be investigated by linearizing the equations for a shallow atmosphere (5.16), (5.19), and (5.20) or the simplified version of these equations (5.24) and (5.25), in the geostrophic approximation. Linearization and the subsequent transformations yield the following equation for the perturbations:

$$p_1'' + \left[\frac{p_0' - p_0''' - \gamma p_0'}{c + p_0'} + \frac{(\gamma - 1)\rho_0' + u_R}{c + \sigma p_0'} - 1 - k^2\right]p_1 = 0. \tag{5.32}$$

Here pressure perturbations are sought in the form $P_1(y)\exp[ik(x - ct)]$, with $\sigma \equiv (\gamma - 1)/\gamma$. The velocity of the zonal flow equals $-p_0'$ (in dimensionless units).

In the limit $\gamma \to \infty$ this equation becomes the equation for an incompressible fluid with an inhomogeneous density profile.

$$p_1'' - \left[\frac{q'}{c + p_0'} + 1 + k^2\right]p_1 = 0, \tag{5.33}$$

where $q = \nabla^2 p_0 - p_0 - \rho_0 - u_R y$ is the so-called generalized flow vorticity. By analogy with Rayleigh's theorem, proceeding as in the derivation of (5.31), we obtain the result that a sufficient condition for stability of an incompressible rotating fluid is the inequality

$$p_0' - p_0''' - \rho_0' + u_R \neq 0. \qquad (5.34)$$

In other words, the generalized vorticity of a stable zonal flow must not have an extremum. When the Rossby radius tends to infinity and $\rho_0^1 = 0$, Rayleigh's theorem follows from (5.34). The Kuo theorem (Lin, 1955), which also follows from (5.34), is a generalization of Rayleigh's theorem to the case of rotation (finite Rossby radius) and $\rho_0' = 0$. In the more general case of a compressible atmosphere with an inhomogeneous density profile a sufficient condition for stability can be obtained from (5.32)

$$(p_0' - p_0''' - \gamma\rho_0')(u_R + (\gamma - 1)\rho_0') > 0. \qquad (5.35)$$

as with (5.31).

Instability can be proved by constructing the eigenfunctions of the differential equation (5.32), which may yield different results depending on the specific circumstances. Examples of the construction of such eigenfunctions in the quasi-classical approximation that includes the Stokes effect are given in the review by Timofeev (1971)

5.6. LARICHEV–REZNIK SOLITON SOLUTIONS AND VORTEX STREETS

The Charney equation and the equations for a shallow atmosphere are remarkable in that, in contrast to equations like the Korteweg–de Vries equation, they involve nonlinearities in the form of a Jacobian, which vanish in the 1D or axisymmetrical cases. Nonlinearities of this sort are typical of waves in gyrotropic media such as rotating atmospheres, magnetized plasmas, etc. It is noteworthy that the Korteweg–de Vries equation was formulated a century ago, while its general solutions were obtained using the inverse scattering method only in the late 1960s. This was done by reducing the Korteweg–de Vries equation to a set of linear equations. The Charney equation, which was proposed in 1947, has had a similar history. Only in 1976 was this equation found to have soliton solutions. Since then it found increasingly wider applications in the theory of synoptic vortices in meteorology and oceanography. As with the Korteweg–de Vries equation, the procedure for solving the Charney equation reduces to a linear problem, but, unfortunately, only in the steady-state case.

The simplest travelling-wave solution to the Charney equation is a plane harmonic wave of the form

$$p_0 = A \sin (\mathbf{k}\cdot\mathbf{r} - \omega_{\mathbf{k}}t), \quad \omega_{\mathbf{k}} = \frac{k_x u_R}{1 + k^2}. \tag{5.36}$$

However, this solution is unstable and decays into two waves, the sum of whose frequencies equals $\omega_{\mathbf{k}}$, since the spectrum of Rossby waves satisfies the decay conditions. This means that for any wave with wave vector **k** and frequency $\omega_{\mathbf{k}}$, a wave can be found with wave vector **q** and frequency $\omega_{\mathbf{q}}$ that satisfies the conditions of phase synchronism $\omega_{\mathbf{k}} + \omega_{\mathbf{q}} = \omega_{\mathbf{k}+\mathbf{q}}$. Of these three waves, the one with the highest frequency transmits its energy to the waves with lower frequencies, so that it is attenuated (decays). The growth rate of this process can be found from the following considerations. Each of the waves separately is in a steady state, but when they exist together, they start to affect one another through the nonlinear term. We shall assume that the amplitude of the wave with the highest frequency, $\omega_{\mathbf{k}}$, it so large that it can be considered constant. In such a case the wave amplitudes **q** and **q** + **k** satisfy the equations (only terms with identical frequencies have been retained):

$$\partial_t a_{\mathbf{q}} = V_{\mathbf{k},\mathbf{q}} a_{-\mathbf{k}} a_{\mathbf{q}+\mathbf{k}}, \tag{5.37}$$

$$\partial_t a_{\mathbf{q}+\mathbf{k}} = V_{\mathbf{k},\mathbf{q}} a_{\mathbf{k}} a_{\mathbf{q}}. \tag{5.38}$$

Hence

$$\partial_t^2 a_{\mathbf{q}} = |V_{\mathbf{k},\mathbf{q}}|^2 |a_{\mathbf{k}}|^2 a_{\mathbf{q}}. \tag{5.39}$$

The solution of this equation increases with time at a growth rate $\gamma \simeq |v_{\mathbf{k},\mathbf{q}}||a_{\mathbf{k}}|$. Since any perturbation can be represented as a sum of Fourier harmonics, one deduces the following picture of the nonlinear evolution: each harmonic creates two waves with lower frequencies. Longer wavelengths correspond to lower frequencies. Thus, a perturbation without a compensating source becomes increasingly lower in frequency and larger in scale. As Hasegawa and Mima (1978) have shown, the latitudinal scale grows much more rapidly than the longitudinal scale. As a result, the perturbations tend to form a zonal flow. However, this is not always true: Larichev and Reznik (1976) have shown that the Charney equation has localized solutions in the form of 2D solitons. These solutions are antisymmetric in longitude. Subsequently, more general solutions were found by Flierl *et al.* (1980). It follows that an arbitrary initial perturbation will not spread out completely in the course of time, but part of it will become a random set of solitons. Interaction between solitons of the Larichev–Reznik type has been studied little and then only numerically. Since they are localized exponentially, however, they do not act at a distance. For this reason, a random set of solitons has a much longer lifetime than a set of sinusoidal waves. It is therefore to be expected that the Rossby wave turbulence is structured in nature and consists of a set of solitons. This is partly confirmed by numerical and laboratory experiments. Here we present a solution of the Charney equation in the form of a soliton. We seek a solution of the form $p = p(\xi, y)$, where $\xi = x - ut$ and u is the displacement velocity. Equation (5.26) then reduces to

$$\{q, p + uy\} = 0, \quad q = \nabla^2 p - p - u_R y, \tag{5.40}$$

where q is the generalized vorticity which also shows up in Kuo's theorem. The general solution of (5.40) is $q = f(p + uy)$, where f is an arbitrary function. Following Larichev and Reznik (1976), we introduce polar coordinates with $r^2 = y^2 + \xi^2$ and $\tan\theta = y/\xi$, and a circle of radius a. It is not known whether this equation has a soliton solution when f is an arbitrary function, or whether the solution can be completely smooth. Larichev and Reznik suggested reducing the problem of finding a solution of (5.40) to a piecewise linear one by choosing f as a linear function with different coefficients inside and outside the circle $r = a$. Equation (5.40) is then similar to the Schrödinger equation with a potential in the form of a rectangular well. The only physical difference is that it is the perturbation itself which forms the potential well. Note that this procedure has been used by Hill to construct 3D vortices in ordinary hydrodynamics. The main difference is that Hill's moving vortices decay with distance according to a power law (because there is no dispersion), whereas Larichev and Reznik's solutions decay exponentially. Let us define f so as to make (5.40) assume the form

$$\nabla^2 p = \varkappa^2 p, \quad \varkappa^2 = 1 - u_R/u, \quad r \geq a, \tag{5.41}$$

$$\nabla^2 p = -k^2 p - (k^2 + \varkappa^2) u_y + k^2(1 - g_0) p_0 \quad r \leq a, \tag{5.42}$$

where p_0, k, g_0 are the arbitrary constant.

We want $f(p)$ to be determined at the boundary $r = a$ and the Jacobian in (5.40) to remain finite; this is achieved by setting $(p + uy)_a = p_0 =$ const at the boundary. We then have the solution

$$p = p_0F_0(r) + auF_1(r)\sin\theta. \tag{5.43}$$

Here F_0 and F_1 are functions of r alone and are given by

$$\begin{aligned} F_0 &= g_0[J_0(kr)/J_0 - 1] + 1, \\ F_1 &= (\varkappa/k)^2 J_1(kr)/J_1 - (\varkappa^2 + k^2)r/k^2a, \end{aligned} \quad r \le a, \tag{5.44}$$

and

$$F_0 = K_0(\varkappa r)/K_0, \quad F_1 = -K_1(\varkappa r)/K_1, \quad r \ge a, \tag{5.45}$$

Here J_m, K_n are Bessel and McDonald functions. When these are given without their arguments, the argument is understood to be either ka or $\varkappa a$, respectively. The condition of continuity for ∇p at the boundary $r = a$ yields the dispersion relation

$$-kJ_1K_2 = \varkappa K_1J_2, \tag{5.46}$$

which determines k in terms of $\varkappa$ and the constant

$$g_0 = -J_0K_2/J_2K_0 > 1. \tag{5.47}$$

The approximate solution of (5.46) is:

$$ka \simeq 3.9 + 1.2(\varkappa a)^2/(3.4 + (\varkappa a)^2) \tag{5.48}$$

Since we have the dispersion relation (5.46), three free parameters remain in the solution: u, a, and p_0. If p_0 is unequal to zero, then the vorticity $\nabla^2 p$ has a discontinuity equal to $(\varkappa^2 + k^2 g_0)p_0$ at the boundary $r = a$. The vorticity is continuous when $p_0 = 0$, and the solution is antisymmetric (Larichev and Reznik, 1976). It can be seen from (5.45) that the soliton consists of an antisymmetric dipole part which, as a "carrier", has a "rider" (or "core"), namely, a symmetric part of arbitrary amplitude. The carrier amplitude is determined by the velocity u and the localization length a. A rider cannot exist without a carrier. Note also that the displacement velocity u can be much greater than u_R. Hence Larichev–Reznik solitons are also possible in a homogeneous medium where $u_R = 0$ (for example, in the polar region of a planet). Clearly, the

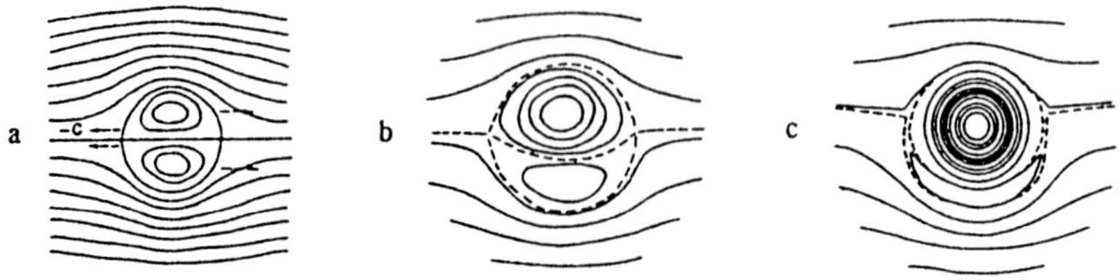

Figure 5.2. Lines of flow in the rest frame of the Larichev–Reznik soliton solution: (a) dipole solution; (b) solution having a small "rider" amplitude (the region of trapped particles coincides with the joining line); (c) solution having a large "rider" amplitude. The joining line is within the trapped particles region.

soliton velocity can have an arbitrary direction in this case. The most interesting feature of the above solutions is the existence of trapped particles which rotate and move together with the soliton. In a soliton of the dipole type the region of trapped particles coincides with the matching region. In the presence of a rider of sufficiently large amplitude, the region of trapped particles can become broader. The region of trapped particles in the rest frame of the soliton is given by the equation $p + uy = 0$. Note that a displacement of the separatrix dividing the region of trapped particles from the matching region makes the soliton more stable, even though there is a discontinuity of vorticity at the matching boundary. Figure 5.2 shows stream lines in the rest frame of the soliton for different rider amplitudes. The presence of trapped particles imparts properties of a vortex to the Larichev–Reznik soliton.

The Charney equation also has a solution in the form of vortex streets in a 2D incompressible fluid. Vortex streets of this sort develop in a zonal flow whose velocity takes on different constant values when $y \to \pm\infty$. To obtain these, we note that (5.40) implies that

$$\nabla^2 p - p - u_R y = f(p + uy) \tag{5.49}$$

where $f(\mu)$ is an arbitrary function.

Suppose the solution is travelling at the Rossby velocity $u = u_R$, and take an arbitrary function of the form

$$f(\mu) = -\mu 4k^2 A \exp(-2\mu/A), \quad \mu = p + uy, \tag{5.50}$$

where A and k are constants whose meaning will be specified below.

In that case (5.49) can be written in the form

$$\nabla^2(p + u_R y) = 4k^2 A \exp\left\{-2(p + u_R y)/A\right\} \tag{5.51}$$

This equation has the solution

$$p + u_R y = A \ln \frac{2}{a}\left[\operatorname{ch} ky + (a^2 - 1)^{1/2} \cos kx\right], \tag{5.52}$$

where a is a parameter that characterizes the amplitude of the vortex street and $2\pi/k$ is the vortex size. When $a = 1$, (5.52) becomes a solution in the form of a zonal flow.

$$p = -u_R y + A \ln 2 \operatorname{ch} ky. \tag{5.53}$$

It is readily verified that this does not fulfil the Kuo criterion which follows from (5.34) for $\rho_0' = 0$. For large values of the argument, (5.53) assumes the form

$$\begin{aligned} p &= (-u_R + kA)y, \quad y \to \infty, \\ p &= (-u_R - kA)y, \quad y \to -\infty. \end{aligned} \tag{5.54}$$

From (5.54) one can see that when $y > 0$, the velocity of zonal flow is equal $u_R - kA$, and when $y < 0$, it is $u_R + kA$. When $a > 1$, there is a vortex street of the sort shown in Figure 5.3 in the middle of the zonal flow. Such solutions are sometimes called "cat's eyes".

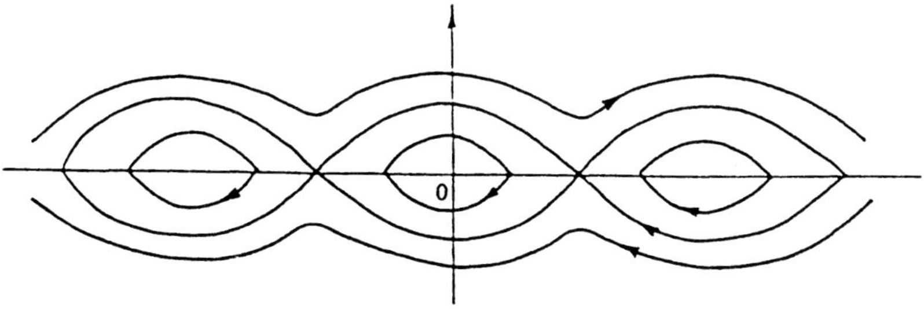

Figure 5.3. The solution of the Charney equation in the form of a zonal flow with a vortex street in the rest frame of the vortices that travel at the Rossby velocity.

We shall now show that the equations for a shallow atmosphere (5.24) and (5.25) also have a solution of the same type as Larichev–Reznik solitons with a vortex street. Let the pressure and density be functions of latitude in such a way that a zonal flow develops. If the typical scale length of the inhomogeneity is large compared to the vortex size, then one can assume approximately that

$$p = 1 + \beta_p y + p_1, \tag{5.55}$$

and

$$\rho = 1 + \beta_\rho y + \rho_1, \tag{5.56}$$

where p_1, ρ_1 is the soliton part of the solution, β_p, β_ρ can be regarded as constant. This corresponds to the presence of a zonal flow at a constant velocity, $-\beta_p$ (in a dimensionless form). According to the criterion (5.35), the distributions of (5.55), (5.56)are not always stable when $p_1 = \rho_1 = 0$. We shall see below that soliton solutions to the equations for a shallow atmosphere exist in both the stable and unstable cases. It is, however, possible that a soliton is not realized in unstable zonal flows. The formation of a vortex street may be more probable in that case. Let a soliton travel at a velocity such that $\partial_t = -u\partial_x$. Then, on substituting (5.55) and (5.56) in (5.24) and (5.25) yield:

$$\{ p_1 + (\beta_p - u)y,\ \nabla^2 p_1 + (\beta_p + u - u_R)y\} = -\{ p_1 + \beta_p y, \rho_1\}, \tag{5.57}$$

and

$$\{ p_1 + uy/\sigma,\quad \rho_1 + u_R y/\gamma\sigma\} = 0, \tag{5.58}$$

where $\sigma \equiv (\gamma - 1)/\gamma$.

It follows from (5.58) that

$$\rho_1 = f(p_1 + uy/\sigma) - u_R y/\gamma\sigma, \tag{5.59}$$

where f is an arbitrary function.

The simplest case is when f is assumed to be linear. Then from the condition that ρ_1 and p_1 in (5.57) are local, we get

$$\rho_1 = (u_R/\gamma u)p_1. \tag{5.60}$$

Substituting (5.60) into (5.57), we obtain an equation in a Jacobian form that can be reduced to the form

$$\nabla^2 p_1 + (u - u_R + \beta_p + \beta_p u_R/\gamma u)y = F(p_1 + (\beta_p - u)y). \tag{5.61}$$

A suitable choice of the arbitrary function in (5.61) can lead to an equation of the form of (5.41) and (5.42), from which the Larichev–Reznik soliton solutions follow. Choosing the arbitrary function of the form (5.50), we obtain a solution in the form of a vortex street of the type (5.52).

5.7. DISPERSIVE SPREADING OF PERTURBATIONS IN GYROTROPIC MEDIA

Perturbations in shallow rotating water are similar to low-frequency perturbations in plasmas localized transverse to a magnetic field. If the frequency of these perturbations is much lower than the Coriolis frequency or the cyclotron frequency, then they cannot propagate in a transverse direction, unless there is some inhomogeneity or nonlinearity. The velocity of propagation in the Larichev–Reznik solutions (5.44) is largely controlled by nonlinear effects. This is a consequence of the fact that these solutions can exist in both inhomogeneous and homogeneous media. The fact that they are exponentially localized indicates their similarity to solitons. However, no direct analogy exists between these and the soliton solutions of the Korteweg–de Vries and Kadomtsev–Petviashvili equations. There is a qualitative difference in that the amplitude δv of the velocity oscillations in the former case is greater than the propagation velocity u ($\alpha \equiv \delta v/u > 1$), whereas in the latter it is much smaller ($\alpha \ll 1$). However, even though $\alpha > 1$, Larichev–Reznik solutions are small perturbations, because the relative amplitude of the pressure perturbation is small in them. This fact figured prominently in the derivation of equations (5.24) through (5.26) in the geostrophic approximation. If we desire a criterion for experimental determination of whether a disturbance is a soliton, we should define the dispersion length (shielding length) l_D and the time of dispersive spreading t_D in shallow, rapidly rotating water. To do this, we rewrite the Charney equation (5.26) in the form

$$\partial_t(\nabla_\perp^2 p - p/r_R^2) - (v_R/r_R^2)\partial_x p = \partial_t\varphi(x - ut, y), \tag{5.62}$$

where the nonlinear term has been replaced by an external force, which has the form of a fixed, exponentially localized function $\partial_t\varphi$.

The solution of the equation has the form

$$p = (\nabla_\perp^2 - r_R^{-2} - v_R/ur_R^2)^{-1}\varphi. \tag{5.63}$$

One can see from this that the shielding length of a travelling localized perturbation is equal to

$$l_D = r_R(1 - v_R/u)^{-1/2}. \tag{5.64}$$

If a disturbance is travelling at a velocity $v_R/u > 1$, it is not shielded. We define the spreading time as the time for the wake of the passing perturbation to disappear. According to (5.63), this is on the order of the length of the wake divided by the velocity of the perturbation, that is,

$$t_D \simeq l_D/u \simeq [(u - v_R)u]^{-1/2} r_R. \tag{5.65}$$

Steady-state vortices can also propagate in shallow water at rest, where $r_R \to \infty$. In contrast to Larichev–Reznik solitons, however, they fall off according to a power law (the velocity of the dipole vortices falls off as r^{-2}), that is, they are not shielded (the shielding length tends to infinity), in agreement with our estimates of l_D and t_D. Thus, if solitons are defined as travelling disturbances that are localized exponentially and do not change during times on the order of t_D, then the Larichev–Reznik solutions can indeed be regarded as solitons. As was remarked in the Introduction, the feature that distinguishes solitons from other types of disturbances is an absence of action-at-a-distance as a result of their strong localization. For the atmosphere, the localization length l_D is given by the planet's rotation and the propagation velocity of the disturbance (see (5.64)). The principal difficulty in predicting atmospheric behaviour is that, apart from other factors, the large-scale, rapidly moving disturbances have a weak correlation between themselves because they are strongly localized.

5.8. STABILITY OF SOLITON SOLUTIONS TO THE CHARNEY EQUATION

One may ask whether the Larichev–Reznik solutions correspond to the observed synoptic vortices. To study this, the relevant solitons must first be investigated for stability. If they are stable, it follows that they are unique. Note that stability has not yet been proven completely for vortices in a non-rotating fluid. The Charney equation (5.26) was previously solved using computers (Flierl *et al.*, 1980; McWilliams and Zabusky, 1982; Larichev and Reznik, 1982) (Figure 5.4). These calculations could

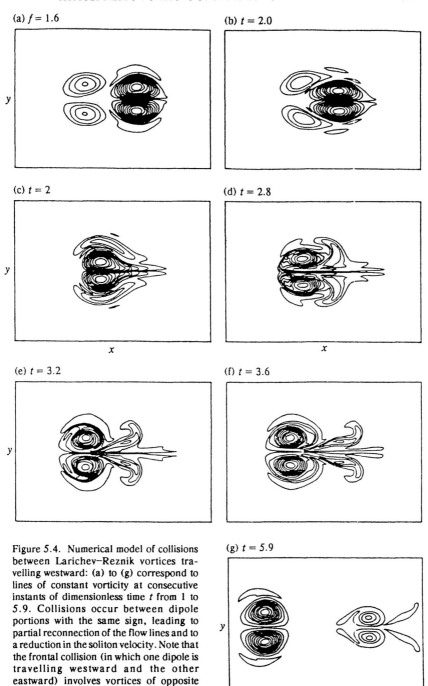

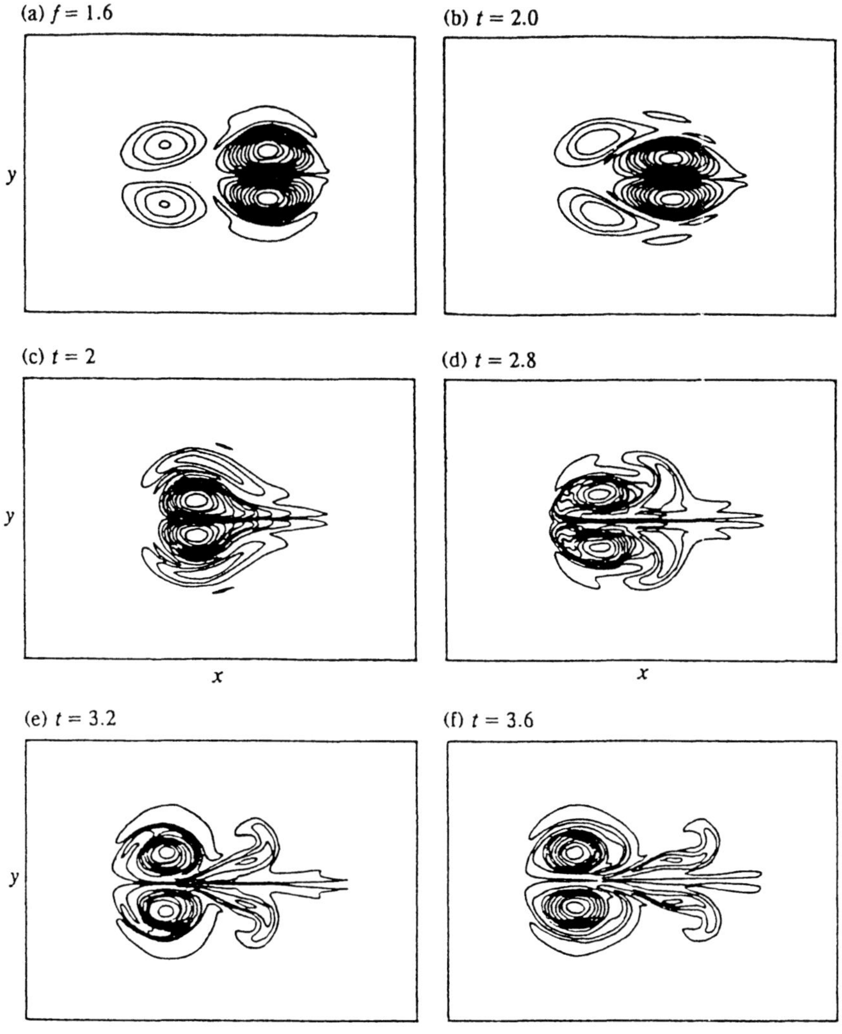

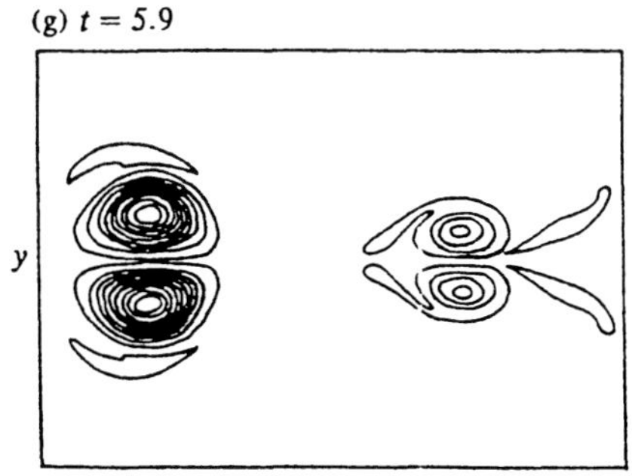

Figure 5.4. Numerical model of collisions between Larichev–Reznik vortices travelling westward: (a) to (g) correspond to lines of constant vorticity at consecutive instants of dimensionless time t from 1 to 5.9. Collisions occur between dipole portions with the same sign, leading to partial reconnection of the flow lines and to a reduction in the soliton velocity. Note that the frontal collision (in which one dipole is travelling westward and the other eastward) involves vortices of opposite signs, hence no reconnection takes place.

not incorporate discontinuities in the vorticity, since a different scheme was used for the integration. Since the Charney equation admits of solutions with a vorticity discontinuity, it cannot be adequately modelled on computers, unless more complex methods are used, for example, the method of test particles. For this reason the existing numerical model of vortices cannot provide compelling evidence to prove their stability, although the models have demonstrated that Larichev–Reznik solitons do not decay through interactions. To prove the stability of soliton solutions to the Charney equation, we make use of the conservation of the following integrals, which are analogues of the integrals for the shallow water equations

$$E = \frac{1}{2}\int \left[(\nabla p)^2 + p^2\right] d^2x,$$

$$N = \int pd^2x,$$

$$Y = \int ypd^2x, \tag{5.66}$$

$$X = \int xpd^2x - u_R tN,$$

$$M = \int \left[g(x - u_R t)^2 + y^2\right] pd^2x,$$

where integrals are taken over the entire fluid volume.

In addition, the integral

$$I_{f,D} = \int_D f(q)d^2x, \quad q \equiv \nabla p - p - u_R y, \tag{5.67}$$

over an arbitrary fluid region is conserved where f is an arbitrary function and D is a fluid region, that is, a region whose points move at a velocity $\zeta \times \nabla p$. Stability can be proved by using the Lyapunov theorem (see Appendix 2). The Lyapunov functional is composed of integrals of motion as follows:

$$-L = -I_{f,i} - I_{g,a} + E + uY - \lambda N, \quad \lambda = \frac{k^2(1 - g_0)}{1 + k^2} p_0, \tag{5.68}$$

where the suffixes a and i denote integration on the outer and inner region of a contour Γ which surrounds D. We put $f = 0.5q^2/(1 + k^2) + f_0$ inside the contour and $g = 0.5u \cdot q^2/u_R + bq$ outside it. Here f_0, k, u, b and λ are constants to be determined. According to the Lyapunov theorem, since L is an integral of motion, those solutions which make the functional L an extremum can be stable, that is those solutions which satisfy the equation $\delta L = 0$. If, in addition, it turns out that $\delta^2 L > 0$, that is, a solution makes L a minimum, then that solution is stable according to Lyapunov.

The condition for L to be an extremum yields Euler's equation in the inner region of Γ in the form

$$\partial_q f = q/(1 + k^2) = -(p + uy) + \lambda. \tag{5.69}$$

Euler's equation in the outer region has the form

$$\partial_q g \equiv qu/u_R + b = -(p + uy) + \lambda. \tag{5.70}$$

The variation should include the variations in both the relative pressure p and Γ. The variation over the contour Γ leads to the conditions $q_i =$ constant and $q_a =$ constant on the contour. The constants are generally different, but they satisfy the condition $f = g$ on the contour Γ. Euler's equations (5.69), (5.70) coincide with the soliton equations (5.41), (5.42). The condition $\partial_q f = \partial_q g$ on the contour yields $(p + uy)_\Gamma = p_0 =$ const.

Since the solution must vanish at infinity in the outer region (5.70), we get $b = \lambda$ and $\varkappa^2 = 1 - u_R/u$. The Larichev–Reznik soliton solution is obtained if Γ is a circle. It is not yet known whether solutions exist for other contour shapes, except that Γ cannot be an ellipse. We now evaluate the second variation of L.

Equation (5.26) can be written as

$$\partial_t q + \mathrm{v} \cdot \nabla q = 0; \quad \mathrm{div}\, \mathrm{v} = 0. \tag{5.71}$$

Arnold (1989) has shown that in this case the regular foliation of q space in "equivalent vorticity" leaves is possible, invariant relative to (5.71). So investigation of $\delta^2 L$ on a leaf of this foliation is sufficient. Because of $\delta L = 0$ the first approximation $\delta q = \xi \cdot \nabla q$ is enough to substitute in L, where ξ is an arbitrary vector with the only restriction $\mathrm{div}\, \xi = 0$. This means that δq is taken tangent to the leaf of the foliation of the phase space in (5.71). Using this we obtain up to quadratic approximation in ξ, after changing the variables of integration:

$$I_{f,i}[q + \delta q] + I_{g,a}[q + \delta q] \approx I_{f,i}[q(\mathrm{x} + \xi)]$$
$$+ I_{g,a}[q(\mathrm{x} + \xi)] = I_{f,i}[q(\mathrm{x})] + I_{g,a}[q(\mathrm{x})].$$

Because of this we have on the leaf where $\delta L = 0$:

$$\delta^2 L = \delta^2 E > 0. \tag{5.72}$$

But the regularity of the foliation can not be proved if there is jump of q on some line and simultaneously $\nabla q = 0$ when approaching to some point of the line. So sufficient condition of stability of solution (5.43) coincides with condition of regularity of foliation and can be written as:

$$\nabla(p + uy)|_{r=a} \neq 0. \tag{5.73}$$

This means that the symmetric part of (5.43) is to be big enough: separatrix points of solution (5.43) must be out of contour Γ (Figure 5.2c). Then (5.43) is stable for any direction of propagation.

The uniqueness of solution (5.43) was proved by Kloeden (1987).

Laedke and Spatchek (1986) have tried to prove the stability of vortices propagating westward by linearization. They demonstrated stability with respect to perturbations in p. They did not, however, include the variation of the contour, so that their proof is uncomplete.

5.9. ANTICYCLONES IN A ZONAL FLOW

Vortices frequently occur in the atmospheres of planets with sizes greater than the Rossby radius. If a vortex is small compared to the inhomogeneity scale length (the planet's radius), then it can be described in the geostrophic approximation. In contrast to Larichev–Reznik vortices, the nonlinearity in the form of a Jacobian then becomes smaller by the ratio of the Rossby radius to the vortex size. Another nonlinearity, one of the Korteweg–de Vries type, becomes important (of the order of the Jacobian) under these conditions. It appears when one includes the dependence of the coefficients of the geostrophic equations on the coordinates and perturbation amplitude (Petviashvili, 1980). When its amplitude is small the velocity of such a vortex is close to the Rossby velocity. The linear terms therefore cancel out and even a small nonlinearity becomes significant. Anticyclone vortices (which rotate in the opposite sense to the planet's rotation) can occur under these conditions. Vortices of this type can be discerned clearly in photographs

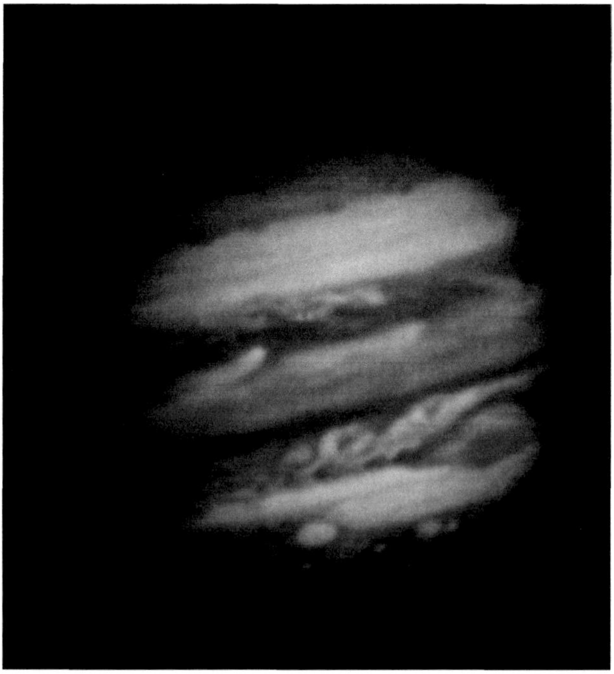

Figure 5.5. Zonal flows and the Great Red Spot in the Jovian atmosphere. The photograph was taken by the Voyager 1 spacecraft.

of Jupiter's atmosphere (Figure 5.5). The Great Red Spot of Jupiter discovered by Robert Hooke as far back as 1664 occupies a special place among these.This spot is located in the southern hemisphere at 22° latitude. Detailed photographs of it were obtained by the spacecrafts Voyager-1,2 in 1979. It lies in a zonal flow with parameters close to the stability threshold defined by (5.34) and (5.35) (Dowling and Ingersoll, 1988). Since the flow vorticity is close to that of the Great Red Spot, the latter is obviously maintained by the zonal flow. Since the Great Red Spot was discovered, it has disappeared and reappeared 17 times. Its colour has also changed many times. The relative pressure disturbance in the Spot is estimated to be 4 to 10 per cent. The vortex core is an ellipse with dimensions of 22 by 11 thousand kilometres. Note that Jupiter's radius is 70 thousand kilometres and its rotation period is 10 hours. The equatorial region has a rotational period that is five minutes shorter than that of the polar region. The spot rotates anticyclonically with a period of about 15 Jovian days and drifts along the latitude, lagging behind the planet's rotation. The drift velocity of the Spot relative to the polar region is 4 to 5 m/s, and that relative to the equatorial region is 120 m/s. The latter value is close to the Rossby velocity at the latitude of the Great Red Spot. In earlier models the Spot was treated as a volcano or as a buoy or a Taylor vortex penetrating deep into the planet. Recently the Great Red Spot has been treated as a Rossby wave soliton (Maxworthy and Redekopp, 1976). There the spot is described by equations that involve a nonlinear term in the form of a Jacobian, while the Korteweg–de Vries nonlinearity is omitted. It will be shown below that this nonlinearity cannot be disregarded, since the dimensions of the Great Red Spot are considerably larger than the Rossby radius, which is 4,000 km at 22° latitude. Petviashvili (1980), (1983) includes the Korteweg–de Vries nonlinearity. The important point was to account for the effect of the zonal flow, which compensates the nonviscous attenuation of the vortex owing to the inhomogeneity in the medium. Petviashvili (1983) shows that in the steady-state case, the parameters of the zonal flow uniquely determine the vortex shape. The stability criterion for zonal flows (5.34) shows that inhomogeneous zonal flows with anticyclonic vorticities become unstable at lower vorticities than cyclone flows. For this reason solitary vortices maintained by zonal flows are anticyclones, while the maintaining zonal flow may be stable. Our starting point is the equation for a generalized vortex frozen in shallow water (5.7); we simplify it using the smallness of ω/Ω, ζ curl v$/\Omega$, and λ/R where λ, ω are the typical size and angular frequency of the perturbations. Then it is sufficient to substitute the expression (5.21) for the velocity in the geostrophic approximation into (5.7). We seek a solution in the geographic coordinates in the form of a steady-state wave travelling latitudinally with an angular velocity u/R, $H = H(\varphi - ut/R, \alpha)$, where φ is the longitude, and α is the latitude. Then we have from (5.7)

$$\frac{u \sin 2\alpha}{2v_0} \partial_\varphi q + \partial_\alpha q \partial_\varphi h - \partial_\alpha h \partial_\varphi q = 0, \tag{5.74}$$

where $v_0 = gH_0/2\omega_o R$, $h \equiv (H - H_0)/H_0$, H_0 is the depth of the unperturbed atmosphere which is considered to be constant for simplicity, R is the planet's radius, $q \simeq [(v_0/S)\nabla^2 h + gS/v_0 R]/(1 + h)$, and $S \equiv \sin \alpha$. The general solution of (5.7) is

$$q = F\left(h - \frac{u(S^2 - S_0^2)}{2v_0}\right), \quad S_0 = \text{const}, \tag{5.75}$$

where F is an arbitrary function. The solution (5.75) is assumed to be a superposition of a zonal flow $h_z(\alpha)$ and the soliton $h_s = h_s(\varphi - ut/R, \alpha)$ which is localized in all directions. It is sufficient to examine (5.75) for latitudes at which the soliton is appreciably different from zero. The shape of the zonal flow and F within that zone are chosen so that (5.75) has a soliton solution. h_z and F outside it may be assumed to be arbitrary functions that smoothly match the solution inside the localization zone of the soliton. Petviashvili (1980) has solved this equation in the absence of zonal flow. The resulting soliton was incompletely localized along the longitude, was incompletely steady-state and subject to nonviscous attenuation. The attenuation can be compensated by maintenance of the soliton by a zonal flow with an appropriate vorticity (Petviashvili, 1983). Assuming that the width of the zone and the soliton amplitude are small enough, it is possible following Petviashvili (1983) to expand into a series, retaining terms that are quadratic in the amplitude and zone width:

$$F(x) = F_0 + F_1 x + \frac{1}{2} F_2 x^2, \tag{5.76}$$

where the F_i are constants to be determined later.

$$x = h - us^2/2v_0 + us_0^2/2v_0, \quad s_0 = \sin \alpha_0, \tag{5.77}$$

and α_0 is the latitude of the zone centre.

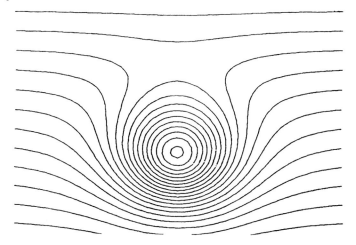

Figure 5.6. Solution of (5.78) in the form of an anticyclone in a sustaining zonal flow. Flow lines in the rest frame of the vortex are shown.

Now we substitute (5.76) into (5.75) and write $h = h_z + h_s$. Then away from the soliton and along the zone, we have an equation for the zonal flow.

$$q\big|_{h=h_z} = F\left(h_z - \frac{u(s^2 - s_0^2)}{2v_0}\right). \tag{5.78}$$

Using (5.75) through (5.78), we get the soliton equation

$$\nabla^2 h_s = \varkappa^2 h_s - A h_s^2, \tag{5.79}$$

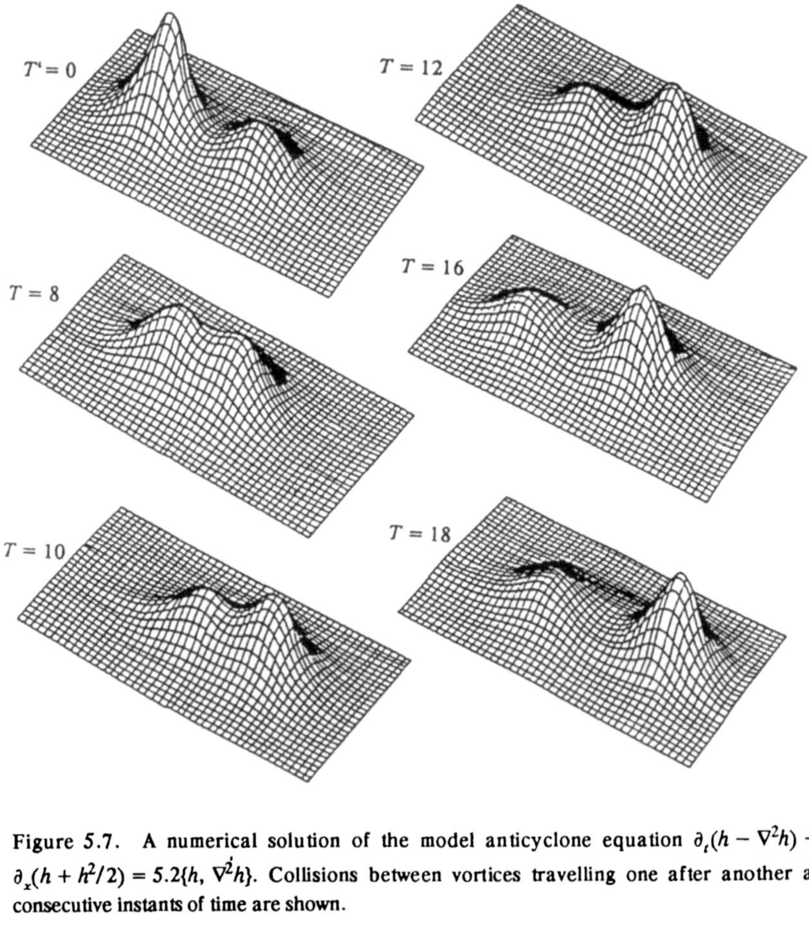

Figure 5.7. A numerical solution of the model anticyclone equation $\partial_t(h - \nabla^2 h) + \partial_x(h + h^2/2) = 5.2\{h, \nabla^2 h\}$. Collisions between vortices travelling one after another at consecutive instants of time are shown.

where $\varkappa^2$ and A are coefficients that depend on α and h_z. Equation (5.79) has a soliton solution satisfying the initial approximations, provided $0 < \varkappa^2 \ll |A|$ (the soliton amplitude is small) and $\varkappa^{-1} \ll R$ (the soliton size is small). The dependence of A on the latitude can be neglected. We also require an extra condition on the dependence of $\varkappa^2$ on the latitude. To illustrate this, consider the 1D equation

$$\partial_x^2 y = \varkappa^2(x)y - y^2. \tag{5.80}$$

Suppose (5.80) has a solitary solution $y(x)$. Then, multiplying this by $\partial_x y$ and integrating over x, we get

$$\int y^2(\partial_x \varkappa^2)dx = 0. \tag{5.81}$$

This can be satisfied if $\varkappa^2$ is an even function of x. The $\varkappa^2$ in (5.78) must satisfy a similar condition. Figure 5.6 shows a soliton in a zonal flow obtained by integrating (5.79) numerically. The stability of this solution is proved by integrating the relevant evolutionary equation. A detailed analysis shows that the only solutions that actually exist are anticyclones ($h > 0$, $A > 0$) (Figure 5.7).

5.10. ROSSBY SOLITONS IN THE DISK OF THE GALAXY

In certain disk galaxies, if one investigates the kinematics of the flat component one finds that noncircular motions may occur in regions outside the visible spiral arms. The departures from circular motion are usually local in character. A good example is NGC, an Sc galaxy whose kinematics have been studied by Zasov and Kyazumov (1981). They point out that in NGC 157, the breakdown of circular motion is associated with a round or oval formation about 4 kpc in diameter; the gas is observed to be moving as a whole relative to the disk rather than expanding. Other examples have been mentioned in the literature.

In the preceding sections of this book it was pointed out that solitons can be supported by the nonuniform zonal flow in rotating planetary atmospheres. Vortices of this kind can also exist in other rotating two-dimensional systems—in particular, as shown by Korchagin and Petviashvili (1985), in the flat components of disk galaxies. These formations will drift as a whole relative to the disk in an azimuthal direction and the motion is supported by the gradient in the angular velocity of the disk as in GRS in zonal flow. It is natural to conjecture that the structures mentioned above are examples of soliton vortices in stable non-uniform rotation of a disk. Perturbations in the flat subsystem of a disk galaxy can be described by the hydrodynamic equations of a two-dimensional fluid. In many cases the disk mass is small enough (compared to the mass of the bulge and the spherical component) for the self-gravitation of the disk to be neglected. Then the equation for the velocity of the matter in the disk will reduce to:

$$d_t \mathbf{v} = -\nabla\varphi - \rho^{-1}\nabla p, \tag{5.82}$$

where φ denotes the gravitational potential of the bulge and the spherical subsystem is assumed to depend on r, z only in a cylindrical coordinate system, ρ is the gas density, and p is the pressure. The gravitational force has vertical and horizontal components. As in the case of 3D synoptical vortices, the vertical component of (5.82) can be simplified in the hydrostatic approximation, neglecting vertical inertial effects:

$$\partial_z\, p/\rho = -\partial_z\, \varphi \equiv g. \tag{5.83}$$

Using this equation and introducing the effective thickness of the disk, H, the horizontal pressure term in (5.82) can be written in the form:

$$\rho^{-1}\nabla_\perp p = \nabla_\perp g H_1, \tag{5.84}$$

where $H_0 = H - H_1$ is the unperturbed part of the effective thickness of the disk.

Let us write the horizontal velocity as a sum $\mathbf{v}_\perp = \mathbf{v}_0(r) + \mathbf{v}_1(r, t)$, where $v_0 = r(\Omega + \omega)$, Ω is the constant part of the angular velocity in the given zone, and $\omega(r)$ is the angular velocity as a function of distance. Using (5.84), we can obtain from (5.82) an equation for the velocity perturbation:

$$(\partial_t + r\,\omega\,\eta\cdot\nabla)\mathbf{v}_1 + (\mathbf{v}_1\cdot\nabla)\mathbf{v}_1 + (\mathbf{v}_1\cdot\nabla)\, r\,\omega\,\eta$$
$$= -\nabla_\perp g H_1 + 2\Omega \mathbf{v}_1 \times \zeta. \tag{5.85}$$

Here ξ, η, ζ are unit vectors along the disk radius, in the azimuthal direction, and along the z axis. The following continuity equation can be introduced for the perturbation H_1 in the thickness

$$(\partial_t + r\,\omega\,\eta\cdot\nabla)H_1 + \operatorname{div} H\mathrm{v}_1 = 0. \tag{5.86}$$

The last two equations are given in a frame of reference rotating with constant angular velocity Ω. This means that the time dependence of the perturbed quantities is taken in a form $H_1(t, r, \alpha - \Omega t)$ and so on, where α is the angle in a cylindrical coordinate system.

We shall regard the characteristic perturbation scale lengths as small compared to the radius of the zone within which we are considering perturbation effects and assume that the disk experiences only weak differential rotation ($\omega \ll \Omega$). As the characteristic time T for a change in the perturbed quantities is assumed greater than the disk rotation period $2\pi/\Omega$, we may express v_1 in terms of H_1 in the geostrophic approximation as

$$\mathrm{v}_1 = \frac{1}{2\Omega}\zeta\times\nabla g H_1 + \mathrm{v}_2, \tag{5.87}$$

where v_2 is the small inertial part of v_1.

On substituting the expression (5.87) into (5.82), dropping small terms, and converting to a dimensionless form, we arrive at the desired model equation:

$$\left(\partial_t + \frac{2\,\omega\,\Omega\,r}{g}\,\partial_y\right)(h - \nabla_\perp^2 h) + v_R\partial_y h = \left\{h, \nabla_\perp^2 h\right\} - r_R(\partial_r g/2g)\partial_y h^2. \tag{5.88}$$

Here the x axis runs along the radius and the y axis along the azimuth; $r_R = c_g/2\Omega$ is the Rossby radius, $c_g^2 = gH_0$, and $v_R = (r_R^2\partial_r^2\omega r - g\partial_r H_0/2\Omega)/c_g$ is the dimensionless drift velocity due to variations in the disk thickness and in the angular velocity. The equation just obtained corresponds to the geostrophic equation (5.7) and differs from the latter in that it contains the operator $(2\,\omega\, r\,\Omega/g)\partial_y$. Note that the last nonlinear term in (5.88) is negligible if the vortex size is on the order of or less than r_R.

As before, (5.88) admits of solutions in the form of steady solitary vortices. Let us seek a solution of the type $h = h(x, y - ut)$ travelling in the azimuthal direction with velocity u (in c_g units). Then we can integrate (5.88) to obtain, as before, the expression

$$q = F\left(h - ux + 2\Omega \int \frac{r\omega}{g}\, dx\right)$$

$$q \equiv \Delta h - h + \int v_R dx - \Omega\left(\partial_x \frac{r\omega}{g}\right) x^2 + \frac{r_R \partial_r g}{2gu} h^2, \qquad (5.89)$$

where F is an arbitrary function. In our case we assume that it is linear. Now we assume that the width of the zone is small compared to R, so that x is also small. Assuming the ordering $x \sim h \sim \omega$ and choosing the proper coefficient in F, we reduce (5.89) to the form (5.79), where

$$\varkappa^2 = -\frac{g}{r\Omega}\, \partial_x \omega; \quad A = -\frac{\partial_x g}{2v_R g}. \qquad (5.90)$$

Soliton solutions exist and are given in Figure 5.6 if $\varkappa^2 > 0$. This means that the angular velocity of the disk decreases with distance from the centre. If the gradient is high, then a linear instability develops and, instead of a soliton, we obtain a vortex street or turbulence may occur.

Finally, let us estimate the characteristic dimensions and amplitude of the soliton solution. Taking $g \simeq 6 \times 10^{-9}$ cm/sec^2, $\partial_r g \simeq -2 \times 10^{-9}$ cm sec^{-2} kpc^{-1}; $H_0 = 500$ pc, and $\Omega = 25$ km sec^{-2} kpc^{-1} for the disk of the Galaxy, we find an Rossby radius of $r_R \simeq 0.6$ kpc. Taking $u \simeq v_R$, we obtain an approximate soliton amplitude of $h \simeq H_1/H_0 \simeq 0.1$. If the soliton had a characteristic scale of $2r_R$, this would correspond to a velocity of vortical rotation of 15 km/sec.

5.11. LABORATORY MODELLING OF ROSSBY SOLITONS

Although observations of water waves have been made for many hundreds of years, they still lead to discoveries of new phenomena and contribute to the understanding of complex processes in different fields of physics. An example is provided by Rossby solitons produced in the laboratory. Petviashvili (1980) suggested modelling large-scale vortical structures in atmospheres and in plasmas by using shallow water in a rotating vessel with a parabolic bottom profile (Figure 5.8). With this shape, the fluid depth is constant for an appropriate velocity of rotation. This is the simplest model for a rotating atmosphere in which the pole of the paraboloid corresponds to the pole of the planet. There is a one-to-one correspondence between the latitudes and longitudes of the paraboloid and the planet, with the eastward direction corresponding to that of the rotation of the paraboloid. The atmosphere itself is concave in the experiment. This model is even more realistic than the relevant theoretical constructions, because it enables one to study many phenomena that are hardly amenable to mathematical description. Since the experiment is so simple, it can serve as a visual aid in teaching the relevant theoretical courses. Because of the similarity (isomorphism) between the mathematical descriptions of Rossby waves and drift waves in plasma discovered by Hasegawa and Mima (1978), experiments with shallow rotating water are of obvious interest for modelling convective transport processes in plasmas as well. Apparatus of this type can be used to simulate flows in atmospheres and oceans, as well as their interaction with vortices and solitons. Earlier experiments with rotating water were carried out under a lid, so that there was no free surface. This led to loss of finite Rossby radius effects.

When conducting these experiments, effects owing to the viscous attenuation of Rossby waves should be kept in mind. They are insignificant in atmospheres and oceans because of the great depths involved. In the laboratory, however, appropriate measures have to be adopted, since the influence of near-bottom viscosity on 2D flows in a rotating fluid is much greater than that for a fluid at rest.

The first paraboloidal vessel for this purpose was built on Petviashvili's initiative in 1981 (Antipov *et al.*, 1981). It was designed to demonstrate the possibility of modelling of Rossby vortices. For this reason the apparatus was comparatively small. The maximum diameter of the vessel was 28 cm and the radius of curvature at the pole was $R = 7$ cm. The period of rotation was 0.6 s. The vessel was intended for observations of vortices on the lateral side of the paraboloid where the Rossby velocity is comparatively large because the Coriolis force is inhomogeneous. Perturbations were produced in the vessel by a rotating disk with dimensions on the order of the Rossby radius. As a result, a vortex was generated, separated from the source, and began drifting westward at a velocity close to (somewhat below) the Rossby velocity. The disk was switched off when the vortex had been created. The evolution of the perturbation was observed by a photographic camera that rotated with the vessel. Small pieces of paper were scattered on the fluid surface to reveal a vortex against the background of the fluid which was at rest

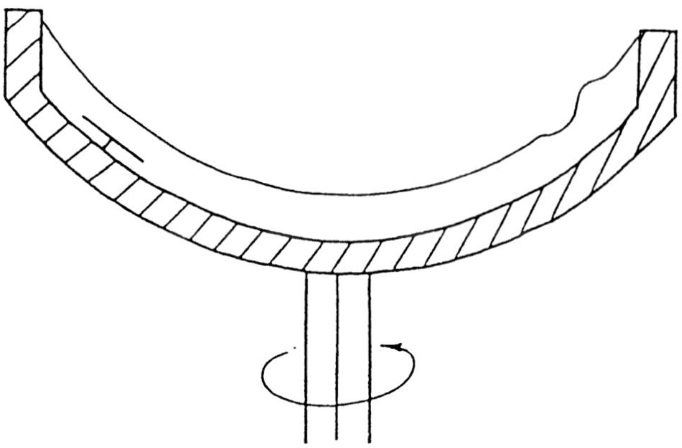

Figure 5.8. A sketch of the cross-section of a vessel designed to model synoptic vortices. When the rotation occurs at a definite rate, the fluid in the vessel has constant depth. Varying the rotation rate will create a gradient in the depth that affects the propagation velocity of vortices. Vortices can be excited by the rotating disk shown on the extreme left. A vertical cross-section of a Larichev–Reznik vortex is shown on the right. The fluid depth must be smaller than the Rossby radius. However, given that the Coriolis force dramatically enhances viscous attenuation, the depth must not be too small.

relative to the vessel. The exposure time was chosen so that one could determine the velocity from trace lengths on the film. Diagnostics with fluid colouring were also used. These experiments produced anticyclone vortices only. Cyclones were swiftly attenuated. The amplitude of the long-lived vortices was rather large, so that the relative perturbation of fluid depth was $h = (H - H_0)/H_0 \simeq 0.5$. The unperturbed fluid depth H_0 was 0.5 cm and the Rossby radius was $r_R \simeq 1.5$ cm. The vortex radius was $(1.5–2)r_R$. The vortex life-time was rather large and claimed to be proportional to H_0^2 (Nezlin, 1986). This error may be ascribed to difficulties in measurements. Subsequently, zonal flows were created using special rings that rotated at the vessel bottom. It was found that when the vorticity in a zonal flow had an anticyclonic direction chains of vortex anticyclones appeared in the zonal flow when it entered an unstable regime (see (5.34)); these were similar to vortices in the absence of zonal flow. When the instability grew (i.e., the gradient of velocity in the zonal flow increased), the vortices became stronger, their dimensions increased, their number decreased and, ultimately, a single powerful vortex remained in the flow.

Since the fluid was very shallow in these experiments, it was feared that bottom friction might have a strong effect on these phenomena. To reduce the role of bottom friction a much larger device was constructed at the Abastumani Astrophysical Observatory (Antonova *et al.*, 1983). In it the operating water depth could be varied from 1.5 cm to 4.5 cm. In contrast to the experiments reported by Antipov *et al.* (1981), this vessel was gently sloping. The radius of curvature was 93 cm with a maximum diameter of 86 cm. The period of rotation was about 1.9 s. The gradient of the unperturbed depth could be changed by varying the rotation period, thus affecting the direction and magnitude of the Rossby velocity. Vortices were generated in a way similar to that used by Antipov *et al.* (1981). These changes in the parameters led to qualitatively novel results. When the depth was $H_0 = 4$ cm, an antisymmetrical cyclone-anticyclone pair was realized as predicted theoretically by Larichev and Reznik (1976) (Figure 5.9). Solitary cyclones which probably correspond to the solution (Flierl *et al.*, 1980) with a vorticity discontinuity (Figure 5.10), were also a frequent occurrence. Solitary anticyclones were also easy to excite (Figure 5.11).

The radii of the vortices were nearly r_R. The larger vortices decayed into smaller ones. Cyclones and anticyclones travelled along latitudes at velocities that were sometimes several times the Rossby velocity. Sometimes they travelled in the opposite direction to the Rossby velocity, an effect that is allowed by the theory (Larichev and Reznik, 1976; Flierl *et al.*, 1980). The cyclone-anticyclone pairs did not always travel along latitudes. They often travelled across the pole (where the Rossby velocity vanishes) and reached the opposite edge of the vessel with nearly the

Figure 5.9. A dipole vortex in shallow water in a rotating vessel (Antonova *et al.*, 1983) travelling westward at a velocity 1.5 times the Rossby velocity.

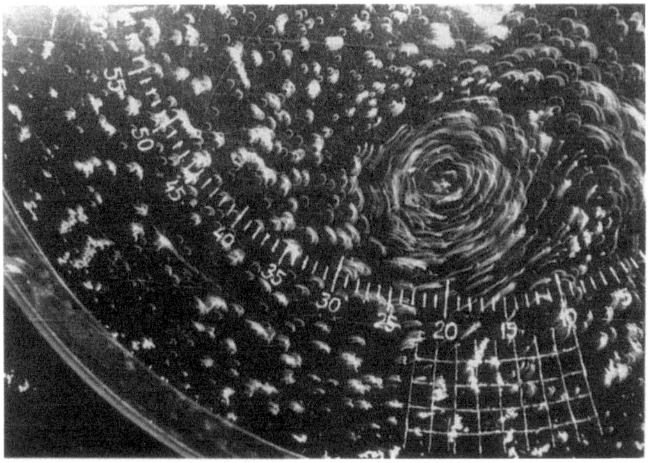

Figure 5.10. A cyclone in shallow water in a parabolic vessel.

Figure 5.11. An anticyclone in shallow water in a parabolic vessel.

same propagation speed. The vortex life-time was close to the Ekman time (the time during which 2D flows are attenuated in the presence of the Coriolis force). We now explain the meaning of this term.

The Coriolis force is known to restrict the thickness l of the bottom layer in a laminar flow. This layer is called the Ekman layer (Pedlosky, 1979). Its thickness is independent of the flow parameters and is given by

$$l = (\nu/\Omega)^{1/2}, \tag{5.91}$$

where ν is the kinematic viscosity ($\nu \simeq 0.01\ \text{cm}^2\ \text{s}^{-1}$ in water) and Ω is the Coriolis parameter. The development of the Ekman layer enhances the dissipation of the energy in the flow. It is in this layer that nearly all attenuation occurs, since the gradient of the velocity in this layer is very large compared to that in the absence of the Coriolis force because of the small thickness of the layer. The typical time for dissipation for a 2D flow can be obtained by multiplying the typical time for dissipation in the layer (l^2/ν) by H_0/l the ratio of the depth of the flow to the thickness of the layer; i.e.,

$$t_E = \frac{l^2}{\nu}\frac{H_0}{l} = \frac{H_0}{(\nu\Omega)^{1/2}}. \tag{5.92}$$

The characteristic time for dissipative attenuation of linear Rossby waves is $t = t_E$ (Pedlosky, 1979). As regards the attenuation of 2D vortices, this problem appears to be more complicated. Two points should be taken into account. The potential energy associated with the variation in the depth inside the vortex increases the lifetime since it is not subject to viscous attenuation. On the other hand, the loss caused by the radial flow near the bottom layer leads to an additional reduction in the lifetime of the vortex because of its finite size. We discuss these two effects separately below.

The characteristic decay time neglecting the effect of radial losses at the bottom can be evaluated from the Charney equation by including Ekman friction:

$$\partial_t(P - \nabla^2 P) + u_R \partial_x P - \{P, \nabla^2 P\} = -t_E^{-1}\nabla^2 P. \tag{5.93}$$

Multiplying this equation by P and integrating, we obtain the damping decrement of the vortex,

$$\gamma_1 \simeq \left[t_E(1 + a^2/r_R^2)\right]^{-1}, \tag{5.94}$$

where a is the characteristic radius of the vortex.

The formula (5.94) shows that the characteristic time for decay of the vortex exceeds the Ekman time. This means that deformation of the free surface makes the vortex less sensitive to dissipation.

The existence of radial flow in the bottom layer decreases the lifetime of the vortex. This effect can be evaluated in the following way. In the bottom layer the radial velocity v_r is on the order of the azimuthal velocity

$v_r \simeq v_\varphi$. The excess volume owing to the surface deformation into the vortex (which is the source of the potential energy) is of order

$$V \simeq \pi a^2 sH \simeq \pi a^3 \Omega v_\varphi / g. \tag{5.95}$$

Here the depth of the surface deformation δH is evaluated from the condition of geostrophic equilibrium. Taking into account the thickness of the bottom layer (5.91), the damping decrement for the vortex owing to the radial loss can be approximated by

$$\gamma_2 \simeq v_r 2\pi al/V \simeq \frac{2}{t_E}\left(\frac{r_R}{a}\right)^2 . \tag{5.96}$$

Combining both these effects, we find the overall characteristic decay time to be

$$t = \frac{1}{\gamma_1 + \gamma_2} = t_E(1 + a^2/r_R^2)\frac{a^2}{r_R^2}(2 + 3a^2/r_R^2)^{-1} . \tag{5.97}$$

From this it follows that vortex decay time is of the order of Ekman time, if the vortex is not too large. In the experiments with shallow rotating water, the observed vortices had radii in the rather narrow range (1–1.5)r_R. Under these conditions, the characteristic vortex decay time according to (5.87) is less than or on the order of t_E.

The lifetime was also observed to depend significantly on the gradient along the depth and the type of vortices. The lifetime of anticyclones did not diminish, in spite of what one should have expected from (5.92). When $H_0 < 2.5$ cm, anticyclones of large amplitude had lifetimes $\gg t_E$. Vortices of all the other types were attenuated almost instantaneously. The long lifetime compared to (5.92) and the relatively large amplitude at $H_0 < 2$ cm lead to the conclusion that anticyclones with a 3D structure (in which vertical motion of the fluid is important) are realized in the regime. The anticyclones in the first sequence of experiments (Antipov *et al.*, 1981), where $H_0 = 0.5$ cm and $h \gtrsim 0.5$, may also have possessed a 3D structure which, even though it differs from atmospheric anticyclones, is of great interest as a new phenomenon that does not fit well into existing theoretical concepts.

Trapped particles were clearly observed in the vortices, as they rotated and travelled together with the vortex. Nobody has succeeded in creating

Rossby solitons without trapped particles. In this regard one may suppose that the trapped particles stabilize the vortices (possibly because closed lines of flow are topologically stable). For particles to be trapped, the velocity of rotation must exceed the velocity of the vortex propagation. Since the latter is greater than the Rossby velocity, the low Rossby velocities are favourable for the creation of weak vortices (those with a small amplitude h). This situation occurs in the polar region of the paraboloid. On the other hand, where the Rossby velocity is high in order for particles to be trapped, the rotational velocity of the vortex must also be high, which causes a considerable perturbation in the depth.

The above experiments have settled the question of whether the observed structures are solitons. We define solitons to be structures in which dispersive spreading is compensated by nonlinear correlation between harmonics. In that case, the lifetime of the observed structures (in case they are solitons) must be large compared to the time for dispersive spreading, t_D (5.64). It should be remembered that this time is not wholly controlled by the Rossby velocity, which may also vanish (for example, in the polar region). It follows from §1.5 that the propagation velocity of packets of linear waves is in the range $v_R/u > 1$. Solitons must not be in resonance with linear waves; this is indeed true for them, because $v_R/u < 1$, where u is the soliton propagation velocity. In other words, the soliton velocity must be different from that of any of the linear waves. The soliton lifetime was large compared to t_D in all the experiments (see (5.65)). However, the vortex velocity was below the Rossby velocity in the experiments in very shallow water (Antipov *et al.*, 1981; Nezlin, 1986). The fact that Rossby waves were not radiated is probably attributable to the fact that the decrement for viscous wave attenuation is very large for small depths (see (5.92)).

We now mention some problems that have not yet been solved in the experiments. Because the depths were shallow, the question of the viscous lifetime of anticyclones has not been completely solved. The cause of the shift in the propagation direction of the cyclone-anticyclone pairs from latitudinal has not been established (whether viscous effects or instability). Because of the small (relatively) lifetime, little has been learned about the interaction of Rossby solitons among themselves and with zonal flows. The generation of zonal flows at large depths is still an open question. Because the existing apparatus operated with small depths, attempts to create a noticeable gradient of temperature have so far been abortive. Such gradients are needed to model the equations for a shallow atmosphere in which isobars and isotherms do not coincide.

Then, as indicated in § 5.6, vortices of new types are possible. One could expect other phenomena that have not been revealed by theoretical analyses. All this points to the need to construct a large vessel with a greater depth of water. This would provide better opportunities to simulate natural conditions. We conclude this section by quoting a few formulas that take into account some peculiarities of a paraboloidal vessel compared with a rotating planet. The surface of a rotating fluid coincides with a paraboloid. In cylindrical coordinates with the z-axis along the axis of rotation and the origin at the pole, the equation for a fluid surface is

$$z = r^2/2R\ , \tag{5.98}$$

where $R = g/\omega_0^2$ is the radius of curvature of the surface at the pole, g the acceleration of gravity, ω_0 the angular velocity.

The vessel bottom is at distance H from this surface. If $H \ll R$, then the water is regarded as shallow, and the bottom shape is close to a paraboloidal shape. The latitude α in a paraboloid is given by

$$\sin \alpha = (1 + r^2/R^2)^{-1/2}, \tag{5.99}$$

where r is the distance from the axis of rotation to the fluid surface. The effective gravity operative in that reference frame is composed of the centrifugal acceleration and the acceleration due to gravity, and is given by $g/\sin \alpha$. It turns out, however, that this circumstance does not affect the expression for the Rossby velocity v_R. Just as in the case of a sphere, it is given by

$$v_R = 0.5H\omega_0(1 + \varkappa \sin \alpha) \cos \alpha, \tag{5.100}$$

where $\varkappa = -d \ln H/d \sin \alpha$ takes the possible dependence of the unperturbed depth H on the latitude into account. We have $v_R = 0$ both at the pole of a paraboloid and on a sphere. We note a slight difference in the Rossby size compared to the formula for a sphere,

$$r_R^2 = gH/\Omega \sin \alpha. \tag{5.101}$$

This is related to the variation on the effective gravity $g_{eff} = g/\sin \alpha$. The Coriolis parameter is given by $\Omega = 2\,\omega_0 \sin \alpha$, as before.

5.12. 3D SOLITONS IN A STRATIFIED ATMOSPHERE

In the preceding we have considered motion in the atmosphere when the horizontal velocity depends weakly on altitude. Since the atmosphere is nonuniform with altitude, flows that are localized in altitude can be generated. The horizontal dispersion length (Rossby radius) played a significant role in horizontal localization. The vertical dispersion length, as will be demonstrated in the following, can lead to vertical localization. As the horizontal dimension of a localized perturbation is large compared to its depth, the vertical velocity will also be small. Therefore, the vertical component of the equation of motion again reduces to the condition for hydrostatic equilibrium,

$$\partial_z P = -\rho g. \tag{5.102}$$

The motion of the atmosphere at altitudes below 100 km is satisfactorily described by the set of hydrodynamic equations (5.11)–(5.13) for an ideal gas. Let us neglect, as before, the Earth's surface curvature, i.e. we restrict ourselves to disturbance on horizontal scales that are much smaller than the Earth's radius R. In addition to the previously mentioned parameters, some new basic parameters affecting the space-time structure of atmospheric disturbances will appear: the Brunt–Väsäla frequency Ω_B and the reduced scale height of the atmosphere H_B, which are given by

$$\Omega_B \equiv \left(-\frac{g}{\rho}\,\partial_z \rho - \frac{g^2}{c_S^2}\right)^{1/2} \simeq g/c_S \tag{5.103}$$

and

$$H_B \equiv \frac{P}{\rho g} \approx \frac{c_S^2}{g} \approx \frac{c_S}{\Omega_B}. \tag{5.104}$$

For simplicity, Ω_B and H_B will be taken as constant. It can be readily shown that in the Earth's atmosphere $\Omega_B \sim 10^{-2}\,\mathrm{s}^{-1}$, $r_R \simeq (2\times 10^6/\sin\alpha)$ m, $H_B \simeq 10^4$ m, $V_R \simeq (100\cot\alpha)\,\mathrm{m.s}^{-1}$, $\Omega \simeq (10^{-4}\sin\alpha)\,\mathrm{s}^{-1}$, and $\gamma = 1.4$.

We limit ourselves, as before, to studying vortices on synoptic scales described in terms of the geostrophic approximation (§ 5.4). Let us assume that the following dimensionless parameters are small

$$|\mathrm{curl}\ \mathbf{v}|/\Omega, \quad \Omega/\Omega_B, \quad H/H_B, \quad a/r_R$$

$$H/a, \quad P'/P, \quad v_z/v_\perp, \tag{5.105}$$

where H and a are the vertical and horizontal dimensions of the vortex, and P' is the pressure perturbation in the vortex. The vertical velocity is inferred from (5.12), (5.13) and (5.102) to be of the form

$$v_z = -(\partial_t + \mathbf{v}_\perp \cdot \nabla)s/(\partial_z s), \quad s \equiv g\rho P^{-1/\gamma}. \tag{5.106}$$

Substituting (5.102) and (5.106) in (5.13) and using the smallness of the parameters (5.105) we obtain

$$\partial_t H_B^2\, \partial_z^2\, P^{\,1/\gamma} - \mathrm{div}\,(P^{1/\gamma}\mathbf{v}_\perp) = 0 \tag{5.107}$$

Now we switch to dimensionless parameters using the following units: time Ω^{-1}, horizontal dimension r_R, vertical dimension H_B, and pressure perturbation γP. Then, by expanding the set of equations (5.21) and (5.107) in powers of the small parameters (5.105), we can reduce this set of equations to the Charney equation, which expresses the conservation of the generalized vorticity in a medium:

$$\partial_t\, q + \{\psi, q\} = 0; \; q = \nabla^2\psi - v_R\, y\,. \tag{5.108}$$

Here $\psi = P'/(\gamma P)$ is the relative pressure perturbation and ∇^2 is the three-dimensional Laplacian. As before, the braces denote the two-dimensional Jacobian in horizontal coordinates. Note that in comparison with the Charney equation, we have dropped the dispersion term in the expression for the generalized vorticity. This is related to the difficulty of simplifying the hydrodynamic equation for the case of 3D perturbations on scales comparable to or greater than r_R and H_B. Thus, (5.108) is valid for 3D perturbations that are not on a very large scale.

The components of the velocity vector are expressed through ψ as

$$\mathbf{v}_\perp = \nabla\psi \times \zeta, \quad v_z = -\partial_{tz}^2\psi - \{\psi, \partial_z\psi\}. \tag{5.109}$$

It should be noted that the units for the horizontal and vertical velocities and scale lengths differ from each other by a large factor of Ω_B/Ω. In terms of the linear approximation, equation (5.108) describes three-dimensional Rossby waves with a dispersion equation of the form $\omega = -v_R k_x/k^2$. Equation (5.108) can be shown to have the following first integrals of motion

$$\int F(q, z)dx\, dy, \quad \int \psi d^3x, \quad \int (\nabla\psi)^2 d^3x\,, \qquad (5.110)$$

where F is an arbitrary function.

For any localized solution it is necessary that $\int \psi d^3x = 0$. The existence of an infinite set of first integrals depending on z indicates that (5.108) is close to being a completely integrable equation. Therefore its soliton solutions for (5.108) may be expected to be stable.

The solutions of (5.108) in the steady-state case were found (Berestov, 1981; Petviashvili, 1988) as three-dimensional solitons moving eastward at a constant velocity u. In the case of steady-state propagating disturbances $\psi = \psi(x - ut, y, z)$, the equation (5.108) takes the form

$$\{\psi + uy, q\} = 0, \quad q = f(\psi + uy, z), \qquad (5.111)$$

where f is an arbitrary function.

Now we transform to spherical coordinates, r, θ, φ, with a change in the vertical scale by a factor of Ω/Ω_B (Ω_B is assumed constant)

$$\begin{aligned} x - ut &= r \sin\theta \cos\varphi, \\ y &= r \sin\theta \sin\varphi, \\ z &= r\cos\theta, \end{aligned} \qquad (5.112)$$

and select f in the form of linear functions with different coefficients inside and outside a sphere of radius r_0. The expression (5.111) can then be presented as

$$\begin{aligned} \nabla^2\psi + v_R y &= -k^2(\psi + uy) + b \quad \text{at} \quad r < r_0, \\ \nabla^2\psi + v_R y &= \varkappa^2(\psi + uy) \quad \text{at} \quad r > r_0, \end{aligned} \qquad (5.113)$$

where $b, k, \varkappa$ are arbitrary constants. It is also required that the matching condition at the sphere

$$(\psi + uy)|_{r=r_0} = G(\theta), \quad \partial_r\psi|_{r=r_0-0} = \partial_z\psi|_{r=r_0+0}, \tag{5.114}$$

(G is an arbitrary function) be satisfied, i.e. the first argument of f from (5.111) must be constant for a fixed value of z on the matching surface. These conditions make the velocity, the pressure, and the pressure gradient continuous. The vorticity may be discontinuous, depending on θ. Once conditions (5.114) have been satisfied and the requirement for exponential localization of the solution at $r \to \infty$ has been met, the solution of (5.113) takes the form

$$\psi = A_e K_{1/2}(\varkappa r) + C_e K_{3/2}(\varkappa r) \sin\theta \sin\varphi$$

$$+ D_e K_{5/2}(\varkappa r)(3\cos^2\theta - 1)(\varkappa r)^{-1/2}, \quad r > r_0$$

$$\psi = A_i J_{1/2}(k r) + C_i J_{3/2}(k r) \sin\theta \sin\varphi$$

$$+ D_i J_{5/2}(k r)(3\cos^2\theta - 1)(k r)^{-1/2}$$

$$- \frac{v_R(k^2 + \varkappa^2)}{k^2\varkappa^2} r \sin\theta \sin\varphi + \frac{b}{k^2}, \quad r < r_0. \tag{5.115}$$

Here J and K are the Bessel and McDonald functions, respectively. The coefficients in equation (5.115) are defined by the relations

$$A_i = -\frac{b r_0 \varkappa^2 (k r_0)^{1/2}}{3k(k^2 + \varkappa^2)J_{3/2}(k r_0)}; \quad A_e = \frac{b r_0 \varkappa (\varkappa r_0)^{1/2}}{3(k^2 + \varkappa^2)K_{3/2}(\varkappa r_0)};$$

$$D_i = D_e \left(\frac{k}{\varkappa}\right)^{1/2} \frac{K_{5/2}(\varkappa r_0)}{J_{5/2}(k r_0)}; \quad C_e = -\frac{v_R r_0 (\varkappa r_0)^{1/2}}{\varkappa^2 K_{3/2}(\varkappa r_0)};$$

$$C_i = \frac{v_R r_0 (k r_0)^{1/2}}{k^2 J_{3/2}(k r_0)}, \tag{5.116}$$

so as to satisfy the matching conditions (5.114). These solutions are localized exponentially, since the McDonald functions vanish exponentially for large arguments. It should be noted that at half-integer orders, these special functions can be expressed in terms of elementary functions. A dispersion equation relating k to $\varkappa$ and r_0 follows from (5.115):

$$\frac{J_{5/2}(kr_0)}{kr_0 J_{3/2}(kr_0)} = -\frac{K_{5/2}(\varkappa r_0)}{\varkappa r_0 K_{3/2}(\varkappa r_0)}. \tag{5.117}$$

The solution of (5.117) that corresponds to the minimal number of nodes in a soliton can be expressed approximately as

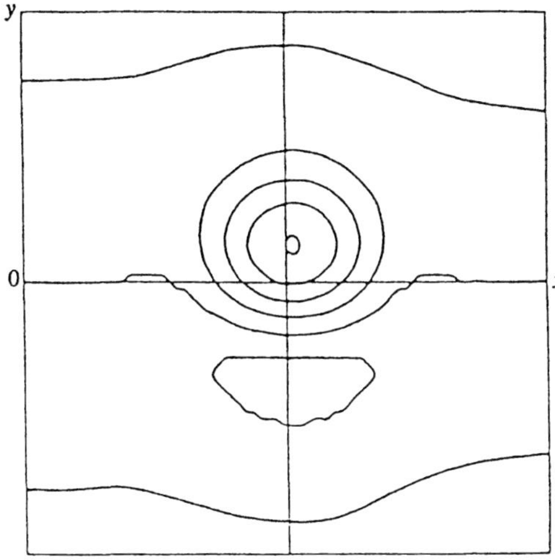

Figure 5.12. The distribution of streamlines in the $z = 0$ plane of a 3D vortex in the atmosphere.

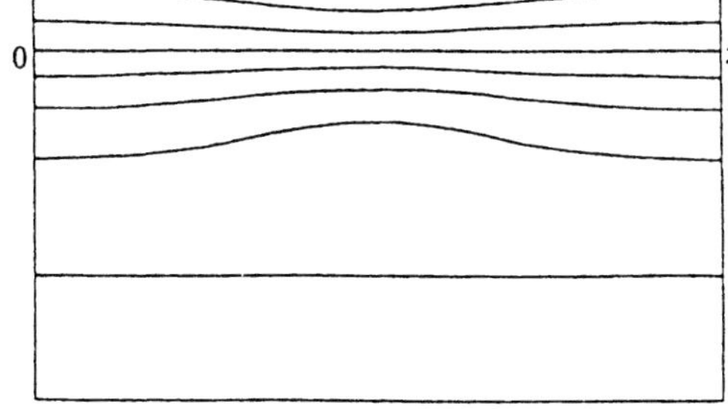

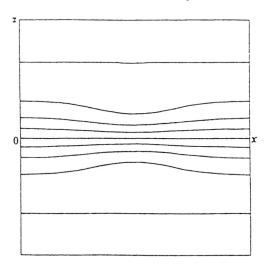

Figure 5.13. The distribution of streamlines in the $y = 0$ plane of a 3D atmospheric vortex. The scale in the vertical direction is nearly one hundred times smaller than the horizontal scale.

$$kr_0 = (4.5(\varkappa r_0)^2 + 86)/((\varkappa r_0)^2 + 15). \qquad (5.118)$$

The propagation velocity of the disturbance

$$u = -v_R/\varkappa^2; \quad v_R > 0, \quad \varkappa \gg 1 \qquad (5.119)$$

follows from (5.113). According to (5.119) soliton solutions move eastward and depend on four free parameters, namely, D_i, b, r_0 and $\varkappa$ (or u), and are a sum of three terms: a monopole vortex (with amplitude proportional to b), a dipole vortex, and quadrupole vortex (with amplitude

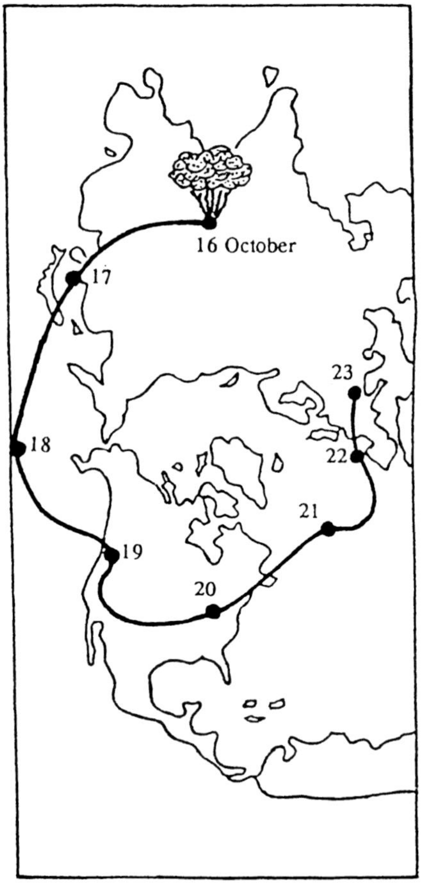

Figure 5.14. The path of debris from a nuclear explosion at 12 km altitude in the atmosphere in 1960. The mean velocity of propagation is 30 m/s. The products may have been trapped in a synoptical vortex created by the explosion.

D_i). By varying b and D_i, we can construct solutions in which one or another term will prevail. We note that Rossby waves propagate westward with a velocity below the Rossby speed.

As an example Figure 5.12 shows the streamlines coinciding with ψ isolines in the $z = 0$ plane and the streamlines in the $y = 0$ plane (Figure 5.13) for $D_i = 0.3$, $r_0 = 0.5$ and $u = 4\times10^{-3}$. It should be noted that the vertical component of the velocity in the symmetry plane of the vortex $z = 0$ vanishes, i.e. the upper and lower half of the vortex are mirror reflections of one another, so that the upper half ($z > 0$) of the vortex may propagate along the solid boundary of the atmosphere, for example along the surface of the Earth or ocean. Impressive evidence of the existence of 3D vortices in the Earth's atmosphere is provided by the observation that the debris from a nuclear explosion propagates with constant velocity at an altitude of 12 km. Figure 5.14 shows that the debris does not diffuse uniformly but stays almost unchanged for many weeks (Lean, 1985).

5.13. THERMAL ENHANCEMENT OF ATMOSPHERIC VORTICES

In the preceding sections it was shown that vortices can be maintained by winds or zonal flows. However in these cases the velocity in the vortices turns out to be close to the wind velocity. For instance, the rotation velocity in the GRS spot is rather close to the zonal flow velocity ($\simeq$ 100 m/s). However, vortices are often observed in which the velocity is much greater than in the background flows. To support such structures some other energy sources are required. In the Earth's atmosphere near the surface of oceans the source of energy can be provided by heat from the condensation of supersaturated vapour. This mechanism can support and augment tropical cyclones. Supersaturated vapour expands with falling pressure and is heated by condensation. Because of the density decrease caused by this process, the buoyancy force increases. This, in turn, leads to an enhanced vapour inflow from the background environment. Part of the energy is transferred to rotation in a horizontal plane and forms the toroidal vortex.

Another enhancement mechanism can occur during sandstorms (simooms) in deserts. Because of the transparency of air, the sand temperature during day time is much higher than that of the background air. Therefore if the heated sand is lifted by a light wind, it will heat the air. As a result, the buoyancy force will appear similar to that in a tropical cyclone. Enhanced vertical movement of the air arising from the buoyancy force will reinforce the draft of heated sand. This may result in the conversion of a light wind into a sand storm. Similar processes may cause the dust storms on the surface of Mars. It should be noted that a coherent stable structure of the wind is necessary in order for such mechanisms to be effective. Random (turbulent) motions cannot be amplified much by such mechanisms. In natural coherent cyclone-like structures of this sort, processes occur which are usually realized in chemical reactors. The simplest example of this is the burning of fuel in an oven.

At higher altitudes the vortices may be enhanced by another mechanism: heat released in exothermic reactions that are sensitive to density and temperature variations and involve ozone (O_3) or the recombination of atomic oxygen O during three-body collisions (Feldstein *et al.*, 1989). It is well known that ultraviolet radiation from the Sun is transformed into chemical energy that builds up in the atmosphere in the form of excessive concentrations of O and O_3 at heights above $\simeq$ 15 km. For this reason, the atmosphere can be regarded as a chemically active medium.

Let us consider the upper mesosphere – lower thermosphere (70–100 km), where the amount of O is greater than the chemical equilibrium value, in more detail. Atomic oxygen disappears in the recombination reactions

$$O + O + M \xrightarrow{k_1} O_2^* + M + \delta E_1$$

and

$$O + O_2 + M \xrightarrow{k_2} O_3 + M + \delta E_2, \qquad (5.120)$$

where O_2^* denotes oxygen in excited states, and M is an arbitrary neutral particle which acquires additional energy as a result of the reaction. The rates of the reactions (5.120) and the energies released in them are $k_1 = 5\times10^{-33}(300/T)^2\,\mathrm{cm^6s^{-1}}$, $k_2 = 6\times10^{-34}(300/T)^2\,\mathrm{cm^6s^{-1}}$, $\delta E_1 = 5.1$ eV, $\delta E_2 = 1.1$ eV respectively, where T is the temperature (K).

For definiteness, assume that the sink for O is entirely through the first reaction (5.120), which is an important source of heat at these altitudes. The continuity equation for the concentration of atomic oxygen, n_O, then takes the form

$$\partial_t n_O + \mathrm{div}\, n_O \mathbf{v} = Q - \Gamma, \quad \Gamma = 2\times10^{-59}\frac{n_O^2 n_M^3}{p^2}, \qquad (5.121)$$

where Q is the source of O owing to solar radiation, Γ/n_O is the recombination rate per unit volume, and n_M is the concentration of M molecules in the atmosphere ($n_O \ll n_M$). We require more exact hydrodynamic equations that account for viscosity and heating. Then instead of (5.11)–(5.13), we obtain

$$d_t \mathrm{v} = -\nabla p/\rho - g\zeta + \Omega(\mathrm{v}\times\zeta) + \mu\Delta\mathrm{v} \tag{5.122}$$

$$\partial_t p + \mathrm{v}\nabla p + \gamma p \,\mathrm{div}\,\mathrm{v} = \delta E_1(\Gamma - Q)\,, \tag{5.123}$$

where μ is the coefficient of viscosity. Heat conduction is neglected in (5.122). For localized disturbances, the conservation of energy takes the following form

$$\partial_t W = \int \left(\frac{\delta E_1(\Gamma - Q)}{(\gamma - 1)} - \mu(\mathrm{curl}\,\mathrm{v})^2\right) d^3x, \tag{5.124}$$

$$W = \int \left(\rho\frac{v^2}{2} + \rho' gz + \frac{p'}{(\gamma - 1)}\right) d^3x, \tag{5.125}$$

where W is the perturbation energy, and a prime denotes the perturbed part of a quantity. Equation (5.124) can be regarded as describing the energy balance of a soliton. Let us show that a perturbation with a finite amplitude ($|\mathrm{div}\,(n_O\mathrm{v})| \gg Q, \Gamma$) can be enhanced. In this case the intensity of the updraft is so high that the variations of temperature and concentration owing to variations in the reaction rates are compensated to a great extent by the influx of "fresh" active matter. Then n_O varies mainly as a result of an upward or downward flow, so that n_O may be treated as proportional to the density of the medium:

$$n_O'/n_O \simeq \rho'/\rho \simeq \partial_z p'/\partial_z p \simeq -\partial_z \psi.$$

As the dissipative terms in (5.122), (5.123) are small, it is possible to substitute the soliton solution in the form (5.115) and (5.116) in these terms. Assuming that $\Gamma = Q$ in an unperturbed atmosphere, we expand $\Gamma - Q$ in series up to the second power. Because $\int\psi d^3x = 0$ and the perturbation is localized, the integrals of the first power terms are equal to zero. Then (5.124) can be written in the form

$$\partial_t W \approx \int \left(10\frac{\delta E_1\Gamma}{(\gamma - 1)}\left(\frac{\partial_z p'}{g\rho}\right)^2 - \frac{\mu v_R^2 r_R^2}{\gamma^2 p^2}(\nabla_\perp \partial_z p')^2\right) d^3x, \tag{5.126}$$

where Γ is given by (5.121). In order for the energy to increase with time, the right-hand side must be positive, which is quite possible for the parameters of the real atmosphere. In this case, the soliton amplitude will be limited by nonlinear terms of higher order which are difficult to evaluate.

On the basis of the distribution of the flow inside a vortex (Figure 5.13), we can conceive the mechanism by which energy is pumped into an anticyclonic vortex from exothermic reactions as follows, limiting ourselves, as an example, to the upper half of the solution (5.115). Oxygen is lost at the vortex centre; the products with a lower value of n_0 move upward and are replaced by fresh active matter supplied from the periphery of the vortex. The temperature at the centre of the vortex appears to rise somewhat, while an atomic oxygen-depleted zone is formed above the centre. Thus, a process usually realized in chemical reactors occurs under natural atmospheric conditions. In the case of a cyclonic vortex, the atmospheric temperature of its centre is lowered because the recombination reaction rate appears to be lower than under the background conditions. It should be noted that the stability of the soliton against minor disturbances is of great importance in processes of this type.

The theory developed above may be used to explain certain events in the lower thermosphere – upper mesosphere, such as the winter anomaly (WA) of the ionospheric D-layer. On WA days electron density n_e increases by one or two orders of magnitude at altitudes of 75–90 km with a peak of 80–85 km at a particular time. The anomalously high n_e occupies a region with characteristic transverse scales of 1000 km and lasts for 5–7 days. It exhibits an irregular structure of WA bursts. Measurements show that in order to explain the WA phenomena we have to find the reasons for temperature rise and, particularly, for the rapid increase in the concentration of NO.

The horizontal scale lengths of WA spots are in good agreement with the transverse dimensions of solitons. In the case of an anticyclone-type soliton in the upper atmosphere, the temperature inside the soliton will increase as a result of intensive burnup of O. Because of the vertical flow in the soliton, a region with a reduced concentration of O forms above it, while an increased NO concentration should be observed inside the soliton. These circumstances suggest that the WA spots of the ionospheric D-layer are related to Rossby solitons.

Rossby solitons may be related to another phenomenon observed at an altitude 90–100 km in the middle latitudes, namely, large-scale regions of the airglow with inhomogeneities in the 557.7 nm line. The horizontal scale of the stratification region is $\simeq$ 1000 km and the vertical scale is $\simeq$ 1 km.

6. Two-dimensional drift vortices in plasmas

6.1. DRIFT WAVES AND CONVECTION IN PLASMAS

It is well known that heat conduction in fluids and gases is predominantly convective in character. A good example of such transport is provided by the formation of Bernard cells in which steady-state vortical fluid motion takes place. Similar phenomena may occur in plasmas confined by a magnetic field. This is confirmed by measurements of heat conduction in experimental devices (Mukhovatov, 1980; Mirnov, 1985; Liewer, 1985).

In the first projects for plasma confinement it was assumed that, if the magnetic field is strong enough, the plasma can move only along the lines of force, that is, its motion is one-dimensional. It was supposed that plasma could easily be isolated from the walls, since the classical diffusion coefficient is proportional to B^{-2}. It seemed that sufficiently strong magnetic fields could easily ensure the necessary conditions for plasma confinement because of a rapid decrease in the losses as B increases. However, it was found out in the late 1950s that plasma can also move across the magnetic field owing to the spontaneous generation of a potential difference φ across the magnetic field. This leads to an $\mathbf{E}\times\mathbf{B}$ drift with a velocity $\mathbf{v}_E = c\,(\mathbf{B}\times\nabla\varphi)/B^2$ and to convective transport of heat and particles.

The theory of convective transport in plasma began with the discovery of drift waves and drift instabilities. The first step in this direction is due to Tserkovnikov (1957). He found low-frequency plasma oscillations travelling across the magnetic field at the diamagnetic drift velocity. The electric field in these was strictly perpendicular to the magnetic field. Rudakov and Sagdeev extended these results to the case of waves whose parallel electric field is small compared to the transverse field. Subsequently this problem has been treated at length (see, for example,

Mikhailovskii, 1974). It has been shown that drift waves can grow larger as a result of collisional dissipation or Landau damping (Timofeev, 1964; Moiseev and Sagdeev, 1964). Estimates of the transport coefficients (Artsimovich and Sagdeev, 1979) based on the approximation of weak turbulence, which assumes that the phases of interacting waves are random, led to abnormally large plasma leakage with a diffusion coefficient on the order of the Bohm diffusion coefficient ($D \propto B^{-1}$). In this connection, mechanisms were sought that could diminish plasma diffusion in drift waves. Galeev (1963) and Mikhailovskaya and Mikhailovskii (1963) have shown that the drift instability can be suppressed by the shear of the confining magnetic field. However, subsequent experiments in machines with large shear still revealed traces of abnormally large plasma conductivity. An attempt was made to overcome these difficulties by developing a nonlinear theory that did not use the random phase hypothesis (Petviashvili and Pokhotelov, 1986). This theory revealed the phenomenon of spontaneous localization in drift waves. It turns out that a qualitative change occurs even at small amplitudes: the waves are self-organized to form a set of solitary vortices. Like solitons, they can exist independently of one another, with some interaction occurring during collisions only. Because of their small dimensions, the vortices are not sensitive to shear. Thus, they can cause abnormally high heat losses, even in machines with large shear. In contrast to turbulent plasma diffusion in drift waves, however, the losses due to vortex formation are convective in character.The quantitative theory of this type of transport is still far from complete.

6.2. PERTURBATIONS OF THE ION DENSITY AND PRESSURE IN DRIFT WAVES

The oscillations of the electron velocity along the magnetic field in drift waves are large compared with the ion velocity. For this reason, oscillations of the ions along B can be neglected. Electrons move across the magnetic field at the electron drift velocity, $\mathbf{v}_E = c(\mathbf{E}\times\mathbf{B})/B^2$. The electric field across $\mathbf{B}$ can be regarded as having a potential. The ion velocity in this direction differs from the electron velocity by a small amount that is related to the comparatively large ion mass. Perturbations in the density and pressure of the ions obey the same equations in all types of drift waves where the wavelength is large compared to the ion Larmor radius. Given this, we now derive equations for the ion density and pressure. Assuming a collisionless plasma, we shall require the kinetic approach for a correct description of the pressure effects. We proceed from the Vlasov equation for the ions in which the parallel electric field and the derivatives with respect to z are neglected under the assumption that they are small quantities of higher order:

$$\partial_t f + \mathbf{v}_\perp \nabla f + (e/m_i)\mathbf{E}_\perp \partial_{\mathbf{v}_\perp} f = \omega_{Bi}\partial_\alpha f, \qquad (6.1)$$

where f is the ion distribution function and α is the angle of the cyclotron rotation in velocity space, i.e., $v_x = v_\perp \cos\alpha$, $v_y = v_\perp \sin\alpha$. Since the coefficients in (6.1) are periodic functions of α with period 2π, the solution is conveniently represented as a Fourier series:

$$f = \sum_k f_k(\mathbf{v}_\perp, v_z, \mathbf{r}, t) \exp(i k \alpha). \qquad (6.2)$$

Substituting this expression into (6.1), we obtain a sequence of equations for the different harmonics. For the first harmonic we have

$$i\omega_{Bi} f_1 = \partial_t f_1 + \langle \mathbf{v}_\perp \nabla_\perp f_0 + (e/m_i)\mathbf{E}_\perp \partial_{\mathbf{v}_\perp} f_0 \rangle_1$$

$$+ \langle \mathbf{v}_\perp \nabla_\perp f_2 \exp(2i\alpha) + (e/m_i)\mathbf{E}_\perp \partial_{\mathbf{v}_\perp} f_2 \exp(2i\alpha)\rangle_1, \qquad (6.3)$$

where $\langle \dots \rangle_k \equiv (2\pi)^{-1}\int_0^{2\pi} d\alpha \exp(-ik\alpha)(\dots)$ is an operator which determines the k-th harmonic amplitude.

We now define a hierarchy of the small parameters involved. We assume $\varepsilon^2 \simeq \omega_{Bi}^{-1} \simeq \partial_t \simeq v_\perp^2 \nabla_\perp^2/\omega_{Bi}^2 \simeq (eE_\perp/m_i\omega_{Bi}v_{Ti})^2 \ll 1$. We shall construct an expansion in powers of the small parameter ε. One can see that the expansion of the k-th harmonic for $k > 0$ starts with a term of order ε^k, so that an expansion in powers of ε: $f_k = f_k^{(k)} + f_k^{(k+1)} + \dots$ is valid, where the superscript denotes the order in ε. We seek f_1 to within terms of order ε^3. To the first order in ε, we have from (6.3) that

$$f_1^{(1)} = -(2\omega_{Bi})^{-1}\Big[v_\perp(\partial_y f_0 + i\partial_x f_0)$$

$$+ (e/m_i)(E_y + iE_x)\partial_{v_\perp} f_0\Big]. \qquad (6.4)$$

We must now calculate f_2 to within ε^2. For it we have the equation

$$2i\omega_{Bi} f^{(2)}{}_2 = \langle \mathbf{v}_\perp \nabla_\perp f^{(1)}{}_1 \exp(i\alpha)$$

$$+ (e/m_i)\mathbf{E}_\perp \partial_{\mathbf{v}_\perp} f^{(1)}{}_1 \exp(i\alpha)\rangle_2. \qquad (6.5)$$

Substituting (6.5) and (6.4) into the right-hand side of (6.3), we obtain the desired expression for $f_1^{(3)}$. (It is easy to see that $f_1^{(2)} = 0$).

Integrating (6.1) over velocity space and using (6.2), we find

$$\partial_t n + \text{div}\,\left(\int \mathbf{v}_\perp f_1 \exp{(i\,\alpha)} d\mathbf{v} + c.c.\right) = 0, \tag{6.6}$$

where $n = \int f_0 d\mathbf{v}$. We also introduce the pressure through the formula $p_i = (m_i/3)\int v^2 f_0 d\mathbf{v}$. Using (6.3), we then obtain the following equation to within terms of order ε^3 from (6.6):

$$\partial_t n + \mathbf{v}_E \nabla n + \text{div}\,(\partial_t + \mathbf{v}_E\cdot\nabla)(en\mathbf{E}_\perp - \nabla p_i)/m_i\,\omega_{Bi}^2 = 0. \tag{6.7}$$

In order to derive an ion pressure equation, one can restrict oneself to the first order in ε, because the pressure appears only in terms that are small in ε.

Multiplying (6.1) by $m_i v^2/3$, we obtain

$$\partial_t p_i + \frac{1}{3}\,\text{div}\left(\int \mathbf{v}_\perp v^2 f_1 d\mathbf{v}\right) - \frac{2}{3}\frac{e}{m_i}\,\mathbf{E}_\perp \int \mathbf{v}_\perp f_1 d\mathbf{v} = 0. \tag{6.8}$$

Substituting f_1 in the first approximation from (6.3) into (6.8), we reduce it to the form

$$(\partial_t + \mathbf{v}_E\cdot\nabla)p_i \equiv d_t p_i = 0. \tag{6.9}$$

The ion contribution to the drift waves is thus described by equations (6.7) and (6.9). The electron density in drift waves is equal to the ion density to a high accuracy. To close the system, therefore, it is enough to have an equation for n in terms of E, which is obtained from the equation for electrons. The drift waves now begin to differ from one another. The principal difference is in the ratio of parallel wave velocity to the Alfvén velocity. If this ratio is large, we have flute waves. If it is of order unity, then we have Alfvén drift waves, where, in contrast to the others, the perturbation of the magnetic field is important. In Kadomtsev's graphic expression, the magnetic field lines lack "stiffness" and are somewhat curved, so that the parallel electric field is reduced. Because of the oscillations in the magnetic field, the electric field becomes a nonpotential field (i.e., is not expressible in terms of an electric potential φ). However, the transverse component of the electric field remains potential and equals $\mathbf{E}_\perp = -\nabla_\perp \varphi$.

The fact that $\mathbf{E}_\perp$ is potential simplifies the problem and allows equations (6.7) and (6.9) to be written compactly. For convenience we shall reduce equations (6.7) and (6.9) to dimensionless form. We define the following dimensionless variables:

$$\omega_{Bi}t \to t,\ r_s^{-1}\mathbf{r}_\perp \to \mathbf{r}_\perp,\ e\varphi/T_{e0} \to \Phi,\ r_s^2 = T_{e0}/m_i,$$
$$n/n_0 \to N + 1,\ p_i/n_0T_{e0} \to p + T_{i0}/T_{e0}. \tag{6.10}$$

The electric potential and pressure are divided by the electron temperature T_{e0} based on considerations that will be clear when we discuss the electron equations. Then (6.7) and (6.9) can, after some simple manipulations, be written in the forms

$$d_t\nabla_\perp^2\Phi = d_t N - \mathrm{div}\{p, \nabla\Phi\}, \tag{6.11}$$

$$d_t p = 0. \tag{6.12}$$

Equation (6.12) has been used in deriving the last term in (6.11). By analogy with the quasigeostrophic equations, we have defined the substantial time derivative $d_t \equiv \partial_t + \mathbf{v}_E\nabla$. Taking into account the expression for the electric drift velocity in dimensionless variables, we have $\mathbf{v}_E\nabla \equiv \{\Phi, \ldots\}$, where the braces denote the Jacobian, as before. For example,

$$\{\Phi, N\} \equiv \partial_x\Phi\partial_y N - \partial_y\Phi\partial_x N. \tag{6.13}$$

These equations can be integrated in the particular case of a steady-state wave packet travelling along y at velocity u. Putting $\partial_t = -u\partial_y$, as well as $d_tP = \{\Phi - ux, P\}$ and so on, from (6.12) we get

$$P = f_p(\Phi - ux), \quad \Phi = \Phi(x, y - ut), \tag{6.14}$$

where f_p is an arbitrary function.

(6.11) can be integrated, if f_p is a linear function of the form

$$f_p = b_p(\Phi - ux)\ , \tag{6.15}$$

where b_p is a constant coefficient.

In that case, (6.11) becomes

$$\{\Phi - u\,x, \nabla^2\Phi - N\} = -b_p\{\Phi - u\,x, \nabla_\perp^2\Phi\}. \tag{6.16}$$

Hence, we easily obtain

$$(1 + b_p)\nabla_\perp^2\Phi = N + f_N(\Phi - u\,x), \tag{6.17}$$

where f_N is an arbitrary function.

Thus, oscillations of the ion pressure P in drift waves (Petviashvili and Pokhotelov, 1986) can significantly affect the steady-state solutions, and in general they should not be identified with density oscillations or disregarded. This is analogous to including baroclinicity in the equations for a shallow atmosphere.

In the other limiting case, where the perturbation size is small compared to the ion Larmor radius, the ions in the potential Φ have a Boltzmann distribution. The ion nonlinearity can then be neglected, as all of the nonlinearity is attributable to the electrons (see § 6.8).

6.3. POTENTIAL DRIFT WAVES

As mentioned in § 1.3, when the angle between the magnetic field and the direction of propagation of the slow ion acoustic wave increases, the phase velocity decreases and can become of the order of the drift velocity. In that case, the ion oscillations become nearly transverse and the density obeys (6.11). As it approaches the drift velocity, the parallel component of phase velocity increases. However, as long as the latter is small compared to the Alfvén velocity, the electric field can be treated as having a potential. If, at the same time, this velocity is also smaller than the thermal electron speed, then the electrons have a Boltzmann distribution in the potential φ:

$$n = n_0(x) \exp\,[e\,\varphi/T_e(x)], \tag{6.18}$$

where the unperturbed plasma density and electron temperature are, generally speaking, fixed functions of x in an inhomogeneous plasma. The system of equations (6.11), (6.12) and (6.18) describes nonlinear electrostatic drift waves. Let us consider its steady-state solution. It has a solution in the form of a solitary vortex and the vortex street. When converted to dimensionless form (6.18) becomes

$$N = N_0(x) \exp\,(\theta\Phi)\,, \tag{6.19}$$

$$N = N_0(x) \exp(\theta\Phi), \tag{6.19}$$

where $N_0(x) = n_0(x)/n_0(0)$ and $\theta(x) \equiv T_e(0)/T_e(x)$.

As noted above, in the steady-state case (6.11) and (6.12) can be reduced to the form (6.17). The coefficients in (6.17), (6.19) depend significantly on x, as in the case of Rossby waves when the perturbation size was large compared with r_R.

We assume that the ratios of the typical scale length of the perturbation, a, to the typical scale lengths of the inhomogeneities in the density $\varkappa_n^{-1}$ and temperature $\varkappa_T^{-1}$ are small. The perturbation is assumed to be localized in the vicinity of $x = 0$ and to be travelling along y. Then the following expansion is valid in the localization zone of the perturbation (Petviashvili, 1983):

$$\begin{gathered} N_0(x) = 1 + \varkappa_n x + \varkappa_n' x^2/2, \\ \theta(x) = 1 - \varkappa_T x, \\ f_N(w) = -1 + f_1 w + f_2 w^2/2. \end{gathered} \tag{6.20}$$

The amplitude Φ is also assumed to be a small quantity of order $\varkappa_n a$. For this reason we can set

$$\exp(\theta\Phi) \simeq 1 + (1 + \varkappa_T x)\Phi + \Phi^2/2. \tag{6.21}$$

Substituting the expansion (6.20), (6.21) into (6.17), (6.19) and discarding terms of higher orders, we obtain

$$\begin{aligned} (1 + b_p)\nabla_\perp^2 \Phi &= (1 + f_1)\Phi + (\varkappa_n - \varkappa_T - f_2 u)\, x\, \Phi \\ &+ (1 + f_2)\Phi^2/2 + (\varkappa_n - u f_1)x + (\varkappa_n' + f_2 u^2/2)x^2. \end{aligned} \tag{6.22}$$

From (6.22) one can see that the following conditions are necessary for soliton solutions to exist:

$$\left.\begin{aligned} \varkappa_n - \varkappa_T - f_2 u &= 0, \\ \varkappa_n - u f_1 &= 0, \\ \varkappa_n' + u^2 f_2/2 &= 0. \end{aligned}\right\} \tag{6.23}$$

Otherwise the solution will not be localized in x. The necessity of the first condition of (6.23) was discussed in § 5.6 in connection with a derivation of anticyclone solutions. Using (6.23), we arrive at an equation for 2D solitons (Petviashvili, 1977):

$$(1 + b_p)\nabla_\perp^2\Phi = \left(1 + \frac{\varkappa_n}{u}\right)\Phi + \left(1 + \frac{\varkappa_n - \varkappa_T}{u}\right)\Phi^2/2. \qquad (6.24)$$

The coefficient b_p is determined from the condition that the ion pressure can be represented as $P = 1 + \varkappa_p x$ in the unperturbed zone. From (6.14) and (6.15), and the requirement that $\Phi \to 0$ outside the zone, we obtain $b_p = -\varkappa_p/u$. In addition, it follows from the requirement that Φ should be small that $1 + \varkappa_n/u \ll \varkappa_T/u$. Finally, we obtain

$$(1 + \tau_i)\nabla_\perp^2\Phi = (1 + \varkappa_n/u)\Phi + \tau_e\Phi^2/2, \qquad (6.25)$$

where $\tau_i \equiv \varkappa_p/\varkappa_n$, $\tau_e \equiv \varkappa_T/\varkappa_n$.

(6.25) includes oscillations in the ion pressure and involves an extra nonlinearity of the Korteweg–de Vries type associated with the gradient of the unperturbed electron temperature. Tasso (1967) was the first to discover this nonlinearity in drift waves. Subsequently Oraevsky, Tasso and Wobig (1969) demonstrated that this nonlinearity, in combination with the dispersion corrections, leads to the appearance of 1D solitons travelling in the direction of the diamagnetic drift. Petviashvili (1977) has used perturbation theory to show that such solitons are unstable and that the equations for the electrostatic drift waves can have 2D soliton solutions of an anticyclone type travelling in the y direction at velocities above the drift velocity ($-\varkappa_n/u < 1$). The solution of (6.25) is plotted in Figure 6.1. In the absence of an electron temperature gradient ($\tau_e = 0$) and neglecting the ion pressure ($\tau_i = 0$), equations (6.11), (6.12) reduce

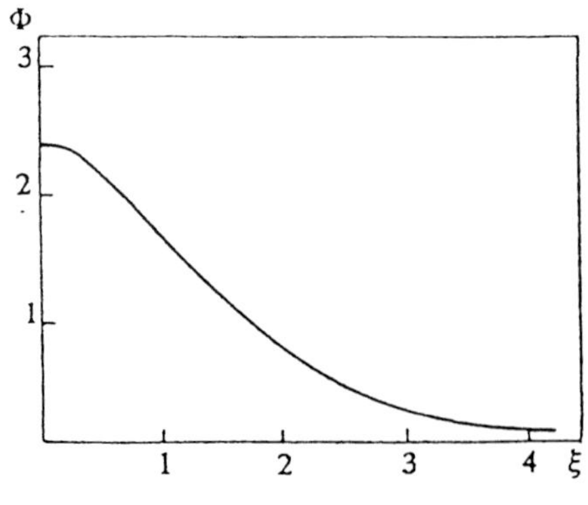

Figure 6.1. Solution of (6.25) in the form of a potential drift soliton of the anticyclone type. The distance from the soliton centre is measured in units of $(1 + \tau_i)/(1 + \varkappa_n/u)^{1/2}$. The soliton amplitude is given in units of $2(1 + \varkappa_n/u)/\tau_e$.

to (5.26), which was derived by Hasegawa and Mima (1978), similarly to the passage from the equations for a shallow atmosphere to the Charney equation. This equation has a solution in the form of Larichev–Reznik solitons. Selecting f_N from (6.17) and (6.19), one can readily obtain solutions in the form of a vortex street on proceeding as in § 5.6.

The electric field in the above solutions is electrostatic, while the ion velocity field is solenoidal. The velocity vorticity is then proportional to $\nabla_\perp^2 \Phi$. Note that the quasineutrality of the oscillations does not break down because the vorticity is finite. This is because quasineutrality breaks down at dimensions on the order of the Debye radius r_D, while finite vorticity affects dimensions of order $r_s \gg r_D$. In a homogeneous plasma with $T_e = T_i$, the size of electrostatic drift vortices is of order r_{Bi}.

Electrostatic drift waves grow in the linear approximation as a result of the dissipative drift instability (Timofeev, 1964). This instability arises because the phase velocity of these waves is in the interval between the electron and ion Larmor drift velocities. For this reason the energy of drift motion is seemingly transferred to the waves. The velocity of the solitons considered above lies outside this interval, so that they are not subject to the instability. They could be maintained by an $\mathbf{E}\times\mathbf{B}$ drift, which is analogous to a zonal flow in the atmosphere. For this to occur, the drift velocity must be inhomogeneous and have a vorticity with an "anticyclone" sign. 2D drift solitons can be maintained in such a flow in a way analogous to anticyclones in the atmosphere as discussed in § 5.9.

6.4. KADOMTSEV–POGUTSE EQUATIONS AND ALFVÉN VORTICES

Alfvén waves occupy a special place in the hydrodynamical theory of stability because they are excited more easily than the other modes in an inhomogeneous plasma. A separate derivation of a simplified equation for Alfvén waves is, therefore, of interest. Nonlinear effects play an important role in the waves, because the velocity of propagation across the magnetic field is small. The effects are extremely small in the 1D case. Thus, the nonlinearity in the Alfvén waves is cubic in the amplitude for longitudinal propagation. When the propagation is nearly perpendicular to $\mathbf{B}_0$, the nonlinearity is quadratic in amplitude, but involves the small parameter ω/ω_{Bi} (Hasegawa and Mima, 1976). Kadomtsev and Pogutse

(1974) have shown that, when the nonlinearity is taken into account, a strong nonlinear term in the form of a Jacobian appears, similarly to the nonlinearity in a 2D fluid. We now derive a simplified equation for nonlinear Alfvén waves. Our starting point is the equations of single-fluid MHD with pressure effects disregarded. The parameters $\partial_z \ll \nabla_\perp$, $\partial_t \ll \omega_{Bi}$ and $B_z - B_0 \ll B_\perp$ are assumed to be small, where B_0 is the unperturbed magnetic field along z. These inequalities yield the result $v_z \ll v_\perp$.

The equation of motion then becomes

$$\rho d_t \mathbf{v}_\perp = (4\pi)^{-1}(\text{curl } \mathbf{B}\times\mathbf{B}), \tag{6.26}$$

$$\partial_t \mathbf{B} = \text{curl } (\mathbf{v}_\perp \times \mathbf{B}), \tag{6.27}$$

and

$$d_t \rho = 0, \quad d_t \equiv \partial_t + \mathbf{v}_\perp \cdot \nabla_\perp. \tag{6.28}$$

We introduce the flux function ψ through the formula

$$\mathbf{v}_\perp = (\zeta\times\nabla\psi), \tag{6.29}$$

and the magnetic potential A. Because $B_z - B_0$ is small, the transverse component of A can be neglected. We then have

$$\mathbf{B}_\perp = (\nabla A_z \times \zeta). \tag{6.30}$$

We now introduce the hierarchy of small parameters $\partial_z/\nabla_\perp \sim \partial_t/c_A\nabla_\perp \sim B_\perp/B_0 \sim \varepsilon$. Taking the curl of (6.26) and using (6.29) and (6.30), we obtain

$$\partial_t \nabla_\perp^2 \psi + \{\psi, \nabla_\perp^2 \psi\} = \{A_z, \nabla_\perp^2 A_z\} - c_A \partial_z \nabla_\perp^2 A_z. \tag{6.31}$$

In this equation density oscillations can be neglected as small quantities of higher order, so we assume $\rho = 1$.

Substituting (6.29) and (6.30) into (6.27), we get an equation relating the vector potential component A_z to the flux function ψ:

$$\partial_t A_z + c_A \partial_z \psi + \{\psi, A_z\} = 0. \tag{6.32}$$

Passing to dimensionless variables from (6.31) and (6.32), we have

$$d_t \nabla_\perp^2 \Phi = \{A, J\} - \partial_z J, \tag{6.33}$$

$$d_t A + \partial_z A = 0, \quad J \equiv \nabla_\perp^2 A. \tag{6.34}$$

Here, we have transformed to the dimensionless units (6.10) and

$$\omega_{Bi} z / c_A \to z, \quad e A_z c_A / c T_e \to A. \tag{6.35}$$

Comparing (6.33) and (6.11), one can identify terms that are due to the electrons through the longitudinal electric current J. One can also see that the flux function is proportional to the electric potential. The left-hand side of (6.34) is equal to $E_{||}$, the longitudinal component of the electric field. The solenoidal solutions of the system (6.33) and (6.34) for $\partial_z = 0$ are called convective cells (Streltsov *et al.*, 1990). In convective cells, the plasma rotates around the magnetic field lines. $A = 0$ there: that is, the magnetic field is not perturbed, while the flux function Φ obeys the equation $d_t \nabla_\perp^2 \Phi = 0$. This system of equations has been used by Kadomtsev and Pogutse (1974) to describe the disruptive instability in tokamaks driven by magnetic reconnection. Stationary vortex solutions of (6.33), (6.34) corresponding to so-called magnetic islands are also well known. These correspond to $\Phi = 0$ and the equation $\partial_z J = \{A, J\}$, which has a solution in the form of the vortex street.

We now seek the more general solutions that travel at velocity u and are inclined at an angle α to the magnetic field. Set $\Phi = \Phi(x, \eta)$, $\eta = y + \alpha z - ut$, and analogously for A. Then from (6.33) and (6.34) we obtain

$$\{\tilde{\Phi}, \tilde{A}\} = 0, \quad \{\tilde{\Phi}, \nabla_\perp^2 \tilde{\Phi}\} = \{\tilde{A}, J\},$$
$$\tilde{A} = A - \alpha x, \quad \tilde{\Phi} = \Phi - u\, x. \tag{6.36}$$

From (6.36) we have

$$\tilde{A} = f_A(\tilde{\Phi}), \quad \nabla_\perp^2 \Phi = f_A' J + f_\Phi(\tilde{\Phi}), \tag{6.37}$$

where f_A and f_Φ are arbitrary functions.

The system (6.36) and (6.37) has a nondenumerable infinity of vortex solutions, because dispersive effects have been neglected in this system.

Here we shall give the simplest of these solutions. Consider a circle of radius a. We select the arbitrary function so that the equality $\Phi = (u/\alpha)A$ is satisfied, as well as

$$\begin{aligned} \nabla_\perp^2 A &= 0, & r \geq a \\ \nabla_\perp^2 A &= -k^2(A - \alpha x - c), & r \leq a \end{aligned} \tag{6.38}$$

where k and c are constants. The solitary solution that falls off at infinity according to a power law has the form

$$A = A_0[J_0(k\,r) - J_0(k\,a)] + \frac{\alpha x}{r}\left[r - 2\frac{J_1(k\,r)}{J_0(k\,a)}\right], \quad r < a, \tag{6.39}$$

$$A = \alpha x(a/r)^2, \quad r \geq a.$$

This satisfies the matching condition $A - \alpha\, x = 0$ at $r = a$, provided $\nabla_\perp^2 A$ is continuous. These requirements lead to the dispersion relation

$$J_1(k\,a) = 0\,. \tag{6.40}$$

In this solution the amplitude A_0, dimension a, velocity of propagation u, and angle of inclination α are arbitrary. The solution consists of a dipole carrier and a monopole core of arbitrary amplitude. A travelling monopole vortex is, thus, always accompanied by a carrier that falls off in accordance with a power law whose parameters are independent of the monopole amplitude. The stability of such solutions is ensured by the fact that the plasma is frozen in the magnetic field (topological stability). Note that, in contrast to linear Alfvén waves travelling at velocity c_A along the magnetic field, the velocity of the above vortices, u/α, can take on arbitrary values. From this one may conclude that such vortices can give rise to anomalous resistivity in an isothermal plasma.

The interaction of vortices was studied in approximation of vortex-current filaments by Petviashvili (1991).

6.5. SOLITARY ALFVÉN VORTICES IN AN INHOMOGENEOUS, FINITE-PRESSURE PLASMA

The preceding section contained a derivation of an equation for Alfvén vortices under the assumption of no dispersion. Petviashvili and Pokhotelov (1985), Kaladze, Petviashvili and Pokhotelov (1986), Kaladze *et al.* (1987) have pointed out that including dispersion along with a density inhomogeneity can lead to new phenomena intrinsic to thermodynamically nonequilibrium systems. It is well known that long-lived fluctuations of large amplitude can occur in such media. For example, sufficiently large drops of a liquid can exist in a supercooled vapour for an indefinitely long time. Inhomogeneous plasmas are also thermodynamically nonequilibrium systems. It is known that linear waves with "negative energies" can exist in such plasmas and their amplitude grows in the presence of dissipation (Kadomtsev, Mikhailovsky and Timofeev, 1965). This instability, however, can be stabilized by the shear of the magnetic field lines. But when the nonlinearity and dispersion are included, drift waves can be localized as 2D vortical tubes of such small dimensions that they would become insensitive to shear. Petviashvili and Pokhotelov (1985), Kaladze *et al.* (1987) have shown this to be the case for Alfvén waves. It has been found that the energy of such vortices can become negative in an inhomogeneous plasma. In other words, the formation of Alfvén vortices is favoured energetically. This phenomenon may be referred to as condensation by analogy with the formation of liquid drops in a supercooled vapour. The formation of such vortices in a laboratory plasma causes convective mixing of particles across the magnetic field, and this is a likely cause of the observed anomalous thermal conductivity. Note that the exponential fall-off in the vortices is related to the inclusion of dispersion in the Alfvén waves (similarly to solitons in dispersive media). If dispersion is not accommodated, the fall-off obeys a power law as shown in the preceding section. Vortex energy with a power-law fall-off diverges in an inhomogeneous plasma, consequently, such solutions are of no interest. In addition, vortices with a power-law fall-off may decay owing to the influence of magnetic shear.

We shall now extend the equations derived in the preceding section to the case where the dispersion of Alfvén waves become important. Consider a plasma of finite pressure $m_e/m_i < \beta \ll 1$. We assume that the typical packet dimension transverse to the magnetic field is within the range $r_s \ll a \ll \varkappa_n^{-1}$ ($\varkappa_n^{-1}$ is the typical size of plasma inhomogeneity).

Transverse perturbations of the magnetic field are again expressed in terms of the z-component of the vector potential through (6.30), while the longitudinal perturbations in **B** can be neglected. The electric field along the magnetic field $\mathbf{B} = \mathbf{B}_\perp + \mathbf{B}_0$ is given by

$$E_\| = -(\mathbf{B}\cdot\nabla\varphi)/B_0 - c^{-1}\partial_t A_z, \tag{6.41}$$

while the transverse component of **E** can be considered potential and equal to $\mathbf{E}_\perp = -\nabla_\perp\varphi$. Nonlinear equations for Alfvén waves in an inhomogeneous plasma can be derived by using the current closure condition

$$\operatorname{div}\mathbf{j} = \operatorname{div}\mathbf{j}_\perp + \operatorname{div}\mathbf{j}_\| = 0. \tag{6.42}$$

In our approximation we have

$$\operatorname{div}\mathbf{j}_\| = -(c/4\pi)(\mathbf{B}\cdot\nabla)\nabla^2(A_z/B_0). \tag{6.43}$$

Only the ions contribute to the divergence of the transverse current. Using the expression for f_1 from § 6.2, to within terms of order ε^3 we obtain

$$\operatorname{div}\mathbf{j}_\perp = cB_0^{-1}\omega_{Bi}^{-1}\operatorname{div}\left\{(\partial_t + \mathbf{v}_E\cdot\nabla)(en_0\mathbf{E} - \nabla p)\right\}. \tag{6.44}$$

When $\beta > m_e/m_i$, the electron inertia can be neglected in the equation for longitudinal electron motion; in such cases it reduces to a balance between the longitudinal electric field and the electron pressure gradient,

$$eE_\| + (\mathbf{B}\cdot\nabla p_e)/n_0B_0 = 0. \tag{6.45}$$

Including the electron pressure p_e leads to dispersion. The contribution of density oscillations in (6.44) and (6.45) can be disregarded, because it gives small corrections of order ω/ω_{Bi}. These corrections are important in the one-dimensional case (Hasegawa and Mima, 1976), when the stronger vectorial nonlinearity vanishes. The latter no longer involves the small parameter ω/ω_{Bi}, so that it is a controlling factor in the two-dimensional case.

This set of equations must be supplemented by the equation of continuity for the electrons

$$(\partial_t + \mathbf{v}_E\cdot\nabla)n - \operatorname{div}(\mathbf{j}_\|/e) = 0, \tag{6.46}$$

and by the equation for the ion pressure (6.9).

It is convenient to transform to dimensionless variables (6.10). We complement these by setting $z\,\omega_{Bi}/C_A \to z$ and $e\,A_z C_A/c\,T_{e0} \to A$. Note that the z coordinate is measured in different units than the coordinates perpendicular to the magnetic field. The velocity along z is measured in units of c_A and that perpendicular to it in units of c_s.

When expressed in these variables, the above equations take the form (Petviashvili and Pokhotelov, 1985):

$$d_t \nabla_\perp^2 \Phi = -d_z J - \operatorname{div}\{p, \nabla_\perp \Phi\}, \tag{6.47}$$

$$\partial_t A = d_z(N - \Phi), \quad J \equiv \nabla_\perp^2 A, \tag{6.48}$$

and

$$d_t N + d_z J = 0, \quad d_t p = 0, \tag{6.49}$$

where $d_z \equiv \partial_z - \{A, \dots\}$, $d_t \equiv \partial_t + \{\Phi, \dots\}$. In the limit of a small perturbation velocity along the magnetic field compared to the Alfvén velocity (that is, when $d_t \ll d_z$), we have $N \to \Phi$ from (6.48) and $A \ll \Phi$ from (6.47). Substituting J from (6.49) into (6.47), we obtain the equation for electrostatic drift waves (6.11). If we put $T_i = 0$ (that is, $P = 0$), we arrive at the Hasegawa–Mima equation (Hasegawa and Mima, 1978; Hasegawa, 1985). Thus, the system (6.47) through (6.49) is universal and describes both drift Alfvén and electrostatic drift waves. It describes a packet localized in a vicinity of the $x = 0$ plane. Far from $x = 0$, we assume that

$$N \to \varkappa_n x, \quad p \to \varkappa_p x, \quad A \to A_0(x), \quad \Phi \to \Phi_0(x). \tag{6.50}$$

For simplicity, $\varkappa_n$ and $\varkappa_p$ will be assumed constant in what follows, and A_0 and Φ_0 will be taken equal to zero (corresponding to the absence of magnetic shear and an ambipolar electric field). The curvature of the unperturbed magnetic field can be incorporated by adding a term of the form $g\,\partial_y\,(N + P)$ (g is the dimensionelss effective gravity) to the left-hand side of (6.47). This effect was taken into account by Petviashvili and Pogutse (1984) in the case of flute perturbations. In the linear approximation, the system (6.47)–(6.49) yields the dispersion relation for drift Alfvén waves (1.38). When transformed to dimensionless units, it becomes

$$\left(1 + \frac{\varkappa_n}{\omega}\right)\left[1 - \frac{\omega^2}{k_z^2}\left(1 - \frac{\varkappa_p}{\omega}\right)\right] + k_\perp^2\left(1 - \frac{\varkappa_p}{\omega}\right) = 0. \tag{6.51}$$

It is convenient to define the perturbed density $N_1 = N - \varkappa_n x$. Then, using

$$\partial_t \int N_1 d^3x = 0. \tag{6.52}$$

We readily verify that the system (6.47)-(6.49) has the energy integral

$$E = \int \left[(\nabla_\perp \Phi)^2 + (\nabla_\perp A)^2 + N_1^2 + 2\varkappa_n x N_1 \right] d^3x. \tag{6.53}$$

The last term in (6.53) corresponds to "energy exchange" between the wave and the plasma, and occurs only in an inhomogeneous plasma ($\varkappa_n \neq 0$). It will be shown below that this term can produce vortices with negative energies.

We shall show that equations (6.47)–(6.49) have a 2D steady-state solution that is localized exponentially. We assume all the quantities to be functions of x and $\eta = y + \alpha z - ut$ only, where α is the angle of inclination of the vortex relative to the magnetic field and u is the propagation velocity of the vortex. The system (6.47) through (6.49) can then be reduced to the form

$$\begin{gathered} \nabla_\perp^2 \Phi = [f_e(\widetilde{A}) + f_\Phi(\widetilde{\Phi})]/(1 + b_p), \\ \nabla_\perp^2 A = f_A(\widetilde{A}) - f_e'(\widetilde{A})\widetilde{\Phi}, \quad \widetilde{\Phi} \equiv \Phi - ux, \\ p = b_p \widetilde{\Phi}, \quad N = \widetilde{\Phi} + f_e(\widetilde{A}), \quad \widetilde{A} \equiv A - \alpha x, \end{gathered} \tag{6.54}$$

where b_p is a constant, f are arbitrary functions, and a prime denotes the derivative with respect to the argument. As before, we represent f as piecewise linear functions:

$$f_e = b_e \widetilde{A}, \quad f_A = b_A \widetilde{A}, \quad f_\Phi = b_\Phi \widetilde{\Phi}. \tag{6.55}$$

In that case (6.54) is replaced by

$$\nabla_\perp^2 \Phi = B(b_\Phi \widetilde{\Phi} + b_e \widetilde{A}), \quad B \equiv (1 + b_p)^{-1}, \tag{6.56}$$

$$\nabla_\perp^2 A = b_A \widetilde{A} - b_e \widetilde{\Phi}, \tag{6.57}$$

$$p = b_p \widetilde{\Phi}, \quad N = \widetilde{\Phi} + b_e \widetilde{A}. \tag{6.58}$$

We define a circle of radius a in the polar coordinates $r^2 = x^2 + \eta^2$, tan $\theta = \eta/x$. An usual matching conditions must be satisfied at the circle, as in the case of electrostatic drift vortices (§ 6.3). However, if we also require $\tilde{A}$ and $\tilde{\Phi}$ to vanish at the circle boundary, then, as shown by Petviashvili and Pokhotelov (1985), the fall-off obeys a power law. Such vortices have long range, like vortices in an incompressible fluid, so they can be easily stabilized by shear. Therefore, vortices with exponential localization are of interest. It has been shown by Kaladze, Petviashvili and Pokhotelov (1986) and Kaladze *et al.* (1987) that an exponential fall-off can be ensured by meeting just one of the conditions $\tilde{\Phi}\big|_a = 0$ or $\tilde{A}\big|_a = 0$. The first case yields only dipole solutions with a discontinuity in the longitudinal current (which is proportional to $\nabla_\perp^2 A$) at the boundary. The second case has continuous vorticity and current. Here, it has not been possible to obtain a solution in the form of a superposition of a dipole and a monopole part (like the carrier and the rider in Larichev–Reznik vortices).

We first consider a solution with a discontinuity in the current at a boundary between trapped and untrapped particles. In that case, the coefficients b_p, b_Φ will be assumed constant everywhere, while b_e and b_A will take on different constant values inside and outside the circles. Then as will be shown below, we obtain solutions that would fall off exponentially at infinity. Outside the circle, a solution is sought in the form

$$\begin{aligned} \Phi &= e_1 K_1(sr)\cos\theta, \\ A &= a_1 K_1(sr)\cos\theta, \end{aligned} \quad r \geq a, \tag{6.59}$$

where K_1 is a McDonald function. Since $P \to \varkappa_p x$ and $N \to \varkappa_n x$ at infinity, this fact and (6.55) yield the coefficients

$$\begin{aligned} b_e &= -(\varkappa_n + u)/\alpha, \quad r \geq a, \\ b_p &= -\varkappa_p/u, \quad 0 < r < \infty. \end{aligned} \tag{6.60}$$

From Φ and A being required to be local and from (6.54) and (6.55) it follows that

$$\begin{aligned} b_A &= b_e u/\alpha = -u(\varkappa_n + u)/\alpha^2, \quad r \geq a, \\ b_\Phi &= 1 + \varkappa_n/u, \quad 0 < r < \infty. \end{aligned} \tag{6.61}$$

Substituting (6.59) into (6.54) and using (6.57) and (6.58), we get an algebraic system of equations for determining e_1 and a_2. The solubility condition for this system yields an expression for the exponential fall-off coefficient

$$s^2 = (\varkappa_n + u)[1/(u - \varkappa_p - u/\alpha^2)]. \tag{6.62}$$

The range of admissible values of u is determined by the condition that s^2 be positive (Figure 6.2). We seek solutions of (6.54) for $r \leq a$ in the form

$$\tilde{\Phi} = [e_2 J_1(k_1 r) + e_3 J_1(k_2 r)]\cos\theta,$$
$$\tilde{A} = [a_2 J_1(k_1 r) + a_3 J_1(k_2 r)]\cos\theta, \tag{6.63}$$

where J_1 is a Bessel function of the first kind. Substitution of (6.63) into (6.54) yields a homogeneous system of equations. The solubility condition for this system determines the coefficients b_e and b_A inside the circle in terms of $k_{1,2}$, n_Φ and b_p:

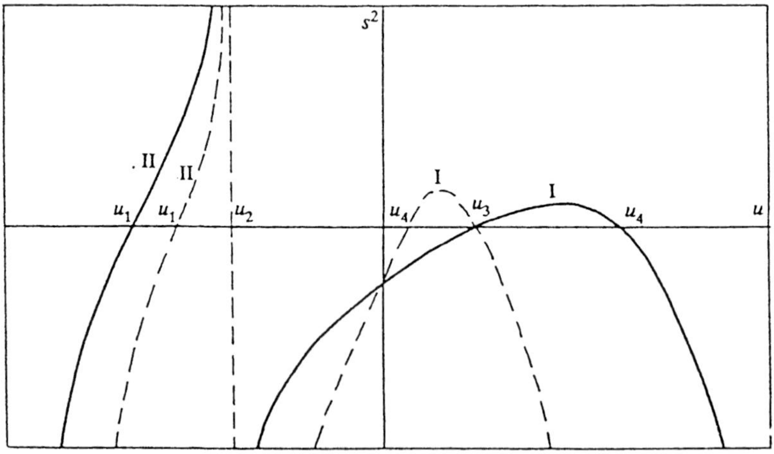

Figure 6.2. The square of the coefficient of exponential decay in Alfvén vortices, s^2, as a function of the velocity u for the case $\beta > m_e/m_i$. The solid line is for the case $\alpha^2 > 2\varkappa_n^2$ and the dashed line for the case $\alpha^2 < 2\varkappa_n^2$ ($\varkappa_n = \varkappa_p$): $u_1 = \varkappa_n - (\varkappa_n^2 + 4\alpha^2)^{1/2}$, $u_2 = \varkappa_n$, $u_3 = -\varkappa_n$, $u_4 = \varkappa_n + (\varkappa_n^2 + 4\alpha^2)^{1/2}$.

$$Bb_e^2 = (k_1^2 + Bb_\Phi)(k_2^2 + Bb_\Phi), \quad r < a, \tag{6.64}$$

$$b_A = -k_1^2 - k_2^2 - Bb_\Phi, \quad r < a, \tag{6.65}$$

where $B = (1 - \varkappa_p/u)^{-1}$ and $Bb_\Phi = (u + \varkappa_n)/(u - \varkappa_p)$. The solutions (6.59) and (6.63) must be matched at the boundary $r = a$. The expressions (6.59) and (6.63) can be regarded as solutions, provided after substitution into (6.54) all the terms are bounded at any point. Discontinuities in the form of finite jumps are allowed. For this to be possible, it is sufficient to require that Φ and $\nabla_\perp \Phi$ be continuous for a choice of the coefficients in the solutions of (6.54) (that is, to require continuity of the density and velocity). However, because b_e and b_A are discontinuous, one must impose a more stringent constraint on A: $\widetilde{A} = 0$ at $r = a$ with continuity of $\nabla_\perp A$ (the perturbed magnetic field). In that case, as can readily be seen from (6.54), $\nabla_\perp \Phi$ is continuous, while the longitudinal current, which is proportional to $\nabla_\perp^2 A$, is discontinuous at the boundary, which is allowed. These requirements determine the coefficients in the solutions (6.59) and (6.63)

$$a_1 = \alpha a / K_1(\mu), \quad e_1 = \alpha a_1/(u - \varkappa_p), \quad r \geq a, \tag{6.66}$$

$$e_2 = \frac{k_1^2 + b_A}{b_e} a_2, \quad e_3 = -\frac{k_2^2 + b_A}{b_e} \frac{J_1(l_1)}{J_1(l_2)}. \tag{6.67}$$

$$a_2 = \frac{\alpha^2 \mu^2 b_e a}{(l_1^2 - l_2^2)(\varkappa_n + u) J_1(l_1)}, \quad a_3 = -\frac{J_1(l_1)}{J_1(l_2)}, \tag{6.68}$$

where $\mu \equiv sa$ and $l_{1,2} = k_{1,2} a$. The quantities α, a, u, and $k_{1,2}$ remain indeterminate for the present.

The matching conditions for $\nabla_\perp A$ and $\nabla_\perp \Phi$ and the requirement $\widetilde{A} = 0$ at the boundary leads to dispersion relations in $k_{1,2}$ that differ from the Larichev–Reznik dispersion relation for quasi-geostrophic vortices (5.46):

$$\frac{a_2 J_1(l_1)}{b_e} \{ l_1^2 g(l_1) - l_2^2 g(l_2) - b_A a^2 [g(l_1) - g(l_2)] \}$$

$$= \frac{\alpha^2 a^3}{u - \varkappa_p}(1 - h) - u a^3, \tag{6.69}$$

and

$$a_2 J_1(l_1)[g(l_1) - g(l_2)] = -\alpha a h, \tag{6.70}$$

where

$$h(\mu) = \mu K_2(\mu)/K_1(\mu) \text{ and } g(l) = 1 - l J_2(l)/J_1(l). \tag{6.71}$$

When the dispersion relations (6.69), (6.70) have been taken into account, the parameters a, $u$and α remain indeterminate. According to the original approximations, the conditions $\varkappa_n \sim \varkappa_p \sim u \sim \alpha \ll 1$ are necessary.

The case $\alpha = u$ corresponds to vortex propagation along $\mathbf{B}_0$ at the Alfvén velocity. The requirement that the vortex size should be large compared to r_s yields the inequalities $a \gg 1$, $k_{1,2}$, $s \ll 1$. Note that α (the angle of propagation in the dimensionless space) corresponds to $\alpha C_s/C_A$. We shall show that the dispersion relation (6.69) and (6.70) have solutions. Note that when l_1 and l_2 are close to zeroes of J_1, the equations become considerably simpler. The theory of Bessel functions gives

$$g(l) = 1 + 2l^2 \sum_{n=1}^{\infty} \frac{1}{l^2 - \gamma_n^2}, \tag{6.72}$$

where the γ_n are the roots of $J_1(\gamma_n) = 0$. Using this formula and the expansion of K_1 for small arguments ($\mu \ll 1$), one easily derives the solution of (6.69) and (6.70) around the poles $|l_1 - \gamma_1| \ll \gamma_1 \simeq 3.8$, $|l_2 - \gamma_2| \ll \gamma_2 \simeq 7.0$ and $|u - u_1| \ll u_1$:

$$\begin{aligned} \frac{\gamma_1}{l_1 - \gamma_1} &\simeq \frac{2a^2}{(u - u_1)(u_1 - u_2)} \\ &\times \left\{ 1 + \left[\frac{u_1 u_2}{a^2} \frac{\gamma_2^2 u_2 - a_2(u_1 + \varkappa_n)}{-\gamma_1^2 u_2 + a^2(u_1 + \varkappa_n)} \right]^{1/2} \right\}, \\ \frac{\gamma_2}{l_2 - \gamma_2} &\simeq \frac{2a^2}{(u - u_1)(u_1 - u_2)} \\ &\times \left\{ 1 + \left[\frac{u_1 u_2}{a^2} \frac{-\gamma_1^2 u_2 + a^2(u_1 + \varkappa_n)}{\gamma_2^2 u_2 - a^2(u_1 + \varkappa_n)} \right]^{1/2} \right\}, \end{aligned} \tag{6.73}$$

where (see Figure 6.2) $u_{1,2} = (1/2)\ [\varkappa_n \pm (\varkappa_n^2 + 4\alpha^2)^{1/2}]$ and $-\varkappa_n < < u < u_1$. Here $\alpha^2 > 2\varkappa_n^2$. This means that the vortices are sub-Alfvénic (their velocity of propagation along $\mathbf{B}_0$ is below the Alfvén velocity). The presence of the roots proves the existence of Alfvén vortex tubes that fall off exponentially at infinity. All the parameters of a tube are expressible in terms of the three free parameters a, u, and α. A tube is a dipole with positive and negative vorticities.

The above formulas readily yield the estimate

$$\Phi \sim A \sim u. \tag{6.74}$$

Hence, we find that the typical rate of rotation v and the magnetic perturbation in the vortex, δB, are on the order of

$$v/C_A \sim \delta B/B_0 \sim \beta^{1/2}\varkappa_n r_s. \tag{6.75}$$

Vortices travelling at the drift velocity $u \simeq -\varkappa_n$ are of particular interest in inhomogeneous plasmas (they can have negative energy, as will be seen below). We have $v \simeq u$ in such vortices.

We now consider a smoother solution in which the longitudinal current is also continuous. We shall assume b_e and b_A to be continuous everywhere, while b_Φ will take on different constant values outside and inside the circle. A solution is again sought in the form (6.56), (6.58). Matching these solutions at the boundary $r = a$, where $\tilde{\Phi} = 0$, we find that the coefficients in (6.56), (6.58) are given by

$$b_p = -\frac{\varkappa_p}{u}, \quad b_e = -\frac{\varkappa_n + u}{\alpha}, \quad b_A = -\frac{u(u+\varkappa_n)}{\alpha^2}, \tag{6.76}$$

$$0 < r < \infty .$$

The coefficient b_Φ outside and inside of the circle takes on the following values

$$b_\Phi = 1 + \varkappa_n/u, \quad r > a, \tag{6.77}$$

$$Bb_\Phi = -\frac{l_1^2}{a^2} - \frac{Bb_e^2 a^2}{l_1^2 + b_A^2 a^2}, \quad r \le a. \tag{6.78}$$

a is given by

$$a^2 = \frac{1}{2s^2}\left[\left((l_1^2 + l_2^2)^2 - \frac{4l_1^2 l_2^2 s^2}{b_A}\right)^{1/2} - (l_1^2 + l_2^2)\right], \quad (6.79)$$

where s is again given by (6.62).

The coefficients in (5.59) and (6.63) are given by

$$e_1 = \frac{ua}{K_1(\mu)}, \quad a_1 = \frac{u - \varkappa_p}{\alpha} e_1,$$

$$e_2 = \frac{u\mu^2 a}{(l_2^2 - l_1^2)J_1(l_1)}, \quad e_3 = -\frac{J_1(l_1)}{J_1(l_2)} e_2, \quad (6.80)$$

$$a_2 = \frac{b_p e_2}{k_1^2 + b_A}, \quad a_3 = -\frac{e_2 b_e}{k_2^2 + b_A}\frac{J_1(l_1)}{J_1(l_2)}.$$

The dispersion relations analogous to (6.69) and (6.70) then have the form

$$\frac{g(l_2) - g(l_1)}{l_2^2 - l_1^2} = \frac{h(\mu)}{\mu^2}, \quad (6.81)$$

$$\frac{l_2^2 g(l_1) - l_1^2 g(l_2) - b_A a^2 [g(l_2) - g(l_1)]}{l_2^2 - l_1^2} = 1 - \frac{b_A a^2 h}{\mu^2}.$$

In contrast to the preceding solution, in which vortical tubes had three free parameters, here the radius a is fixed, so that only two parameters remain free, u and α. In a similar fashion to the preceding, the dispersion relations (6.81) and (6.82) become simpler around zeroes of Bessel functions:

$$l_1 \simeq \gamma_1 (1 + \mu^2/2\gamma_1^2), \quad (6.83)$$

$$l_2 \simeq \gamma_2 (1 + \mu^2/2\gamma_2^2). \quad (6.84)$$

The vortex size a is given by

$$a^2 = 11.3\frac{u_1 - \varkappa_p}{u_1 + \varkappa_n} \gg 1. \quad (6.85)$$

It should be noted that the wavenumbers $k_{1,2}$ were required to be real in the derivation of the solutions in the inner region. This led to some constraints on the coefficients

$$Bb_\Phi + b_A < 0, \quad (b_\Phi b_A - b_e^2) > 0. \tag{6.86}$$

These requirements are readily seen to hold for $-\varkappa_n/u < 1$, which corresponds to sections I and II in Figure 6.2. The relations between the various quantities in this vortex are the same, in terms of order of magnitude, as in a vortex with a discontinuous vorticity. Marnachev (1988) has derived a solution that combines these two. He defines two matching curves which are concentric circles. $\tilde{A}$ = const on one circle and $\tilde{\Phi}$ = const on the other. This solution has as many as four free parameters.

6.6. SOLITARY ALFVÉN VORTICES IN AN INHOMOGENEOUS, LOW-PRESSURE PLASMA

When $\beta < m_e/m_i$, the Alfvén velocity becomes greater than the electron thermal velocity. For this reason, electron pressure fluctuations can be neglected compared to the fluctuations in the kinetic energy of the Alfvén waves (finite electron mass effect). Equations (6.47) through (6.49) are then replaced (Kaladze *et al.*, 1987) by

$$d_t \nabla_\perp^2 A = -d_z J, \tag{6.87}$$

$$\partial_t A - \varepsilon d_t J = d_z(N - \Phi), \tag{6.88}$$

$$d_t N = -d_z J, \quad \varepsilon \equiv m_e/\beta m_i > 1. \tag{6.89}$$

In these equations, the typical dispersion scale length is of order r_{Bi}/ε, which equals the skin depth c/ω_{pe}; the latter is greater than r_s because $\beta < m_e/m_i$. Therefore, ion pressure fluctuations can also be disregarded. In this system an energy integral of the form

$$E = \int [(\nabla_\perp A)^2 + (\nabla_\perp \Phi)^2 + \varepsilon J^2 + N_1^2 + 2\varkappa_n x N_1]\, d^3x \tag{6.90}$$

is conserved. As in the case of a finite-pressure plasma, the Alfvén waves are extremely sensitive to inhomogeneities. For this reason, the last term in (6.90) can sometimes result in a negative energy.

Proceeding in a similar fashion to the preceding section, we derive the equations of the steady-state vortex:

$$\nabla_\perp^2 \Phi = b_\Phi \tilde{\Phi} - b_e \tilde{A}, \tag{6.91}$$

$$J = b_A \tilde{A} - b_e \tilde{\Phi} \tag{6.92}$$

and,

$$N = (1 - \varepsilon b_A)\tilde{\Phi} + b_e \tilde{A}, \tag{6.93}$$

where $\Phi = \Phi(x, \eta)$, $\eta = y + \alpha z - ut$.

We also introduce a circle of radius a. The coefficients b_e and b_A are assumed constant everywhere and are to be determined from the condition that the vortex should vanish at infinity. The coefficient b_Φ is discontinuous. Outside the circle it is found from these conditions and inside the circle from the matching conditions. Since b_Φ is discontinuous at the boundary, we must have the condition $\tilde{\Phi}\big|_{r=a} = 0$. From this and the continuity of A we obtain the following relations:

$$b_e = \frac{\alpha(u + \varkappa_n)}{\varepsilon u^2 - \alpha^2}, \quad b_A = \frac{u(u + \varkappa_n)}{\varepsilon u^2 - \alpha^2}, \quad 0 < r < \infty, \tag{6.94}$$

$$b_\Phi = -\frac{\alpha^2}{u}\frac{u + \varkappa_n}{\varepsilon u^2 - \alpha^2}, \quad r > a. \tag{6.95}$$

The vortex size a and the coefficient b_Φ inside the circle are given by (6.78) and (6.79), as in the case of a finite-pressure plasma.

The solutions of (6.91) and (6.93) coincide formally with those of (6.56) and (6.58). The difference is that, when $\beta < m_e/m_i$, s^2 is given by (Figure 6.3)

$$s^2 = \frac{u + \varkappa_n}{u}\frac{u^2 - \alpha^2}{\varepsilon u^2 - \alpha^2}. \tag{6.96}$$

Matching $\nabla_\perp \Phi$ and $\nabla_\perp A$ at the boundary leads to dispersion relations of the form (6.69) and (6.70) where, however, b_A is given by (6.94). Thus, vortices in the case $\beta < m_e/m_i$ differ from vortices in plasmas with $\beta > m_e/m_i$ only in the expressions for the coefficients (6.94) and (6.95), and in the value of a_1, which is given by

$$a_1 = \frac{ue_1}{\alpha} = \frac{u^2 a}{\alpha K_1(\mu)}. \tag{6.97}$$

In the case of a low-pressure plasma with $\beta < m_e/m_i$, as in the case $\beta > m_e/m_i$, conditions must be satisfied which ensure that solutions of the form (6.63) can be derived in the inner region. In the present case, these conditions are

$$b_\Phi + b_A < 0, \quad b_\Phi b_A - b_e^2 > 0. \tag{6.98}$$

Using (6.78) and (6.94) in a similar fashion to the preceding section, we conclude that the conditions (6.98) hold when

$$\alpha^2 \varepsilon^{1/2} < u < |\varkappa_n|, \tag{6.99}$$

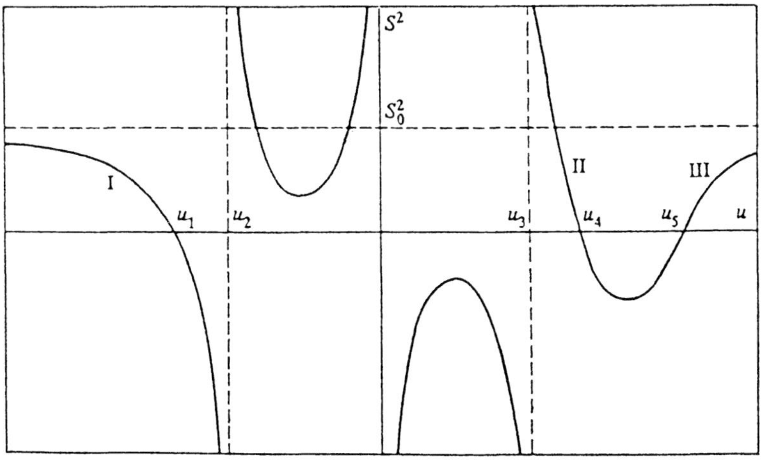

Figure 6.3. The square of the coefficient of exponential decay in vortices, s^2, as a function of the velocity u for the case $\beta < m_e/m_i$: $u_1 = -\alpha$, $u_2 = -\alpha\varepsilon^{-1/2}$, $u_4 = \alpha$, $u_5 = -\varkappa_n$, $s_0^2 = \varepsilon^{-1}$.

that is, in section II of Figure 6.3. Branches I and III in Figure 6.3 correspond to the reverse inequality in (6.98) as $s^2 \to 0$ around u_1 and u_5, respectively. This corresponds to a solution of (6.87)–(6.89) in the inner region of the form

$$\tilde{\Phi} = [e_2 J_1(k_1 r) + e_3 I_1(k_2 r)] \cos\theta,$$
$$\tilde{A} = [a_2 J_1(k_1 r) + a_3 I_1(k_2 r)] \cos\theta, \tag{6.100}$$

where I_1 is a modified Bessel function.

6.7. CONDENSATION OF ALFVÉN WAVES INTO VORTEX TUBES

We now find how the energy of vortex tubes depends on their parameters. Since they are of infinite length, one naturally speaks about their energy per unit length. This has the following form for both low- and finite-pressure plasmas:

$$W = \int \left[(\nabla_\perp A)^2 + (\nabla_\perp \Phi)^2 + \varepsilon J^2 + N_1^2 + 2\varkappa_n x N_1 \right] dx\, d\eta. \tag{6.101}$$

Substituting the solutions obtained above into this expression yields the energy as a function of the tube parameters. In general, this expression is cumbersome. It becomes simpler, however, in the limiting cases considered above. Thus, for vortices with a discontinuous longitudinal current we get

$$W = -4\pi a^2 \varkappa_n^2 / s^2; \quad \varkappa_n^2, s^2 \ll 1, \tag{6.102}$$

while for vortices with a continuous current,

$$W = -\frac{4\pi u(u - \varkappa_p)\varkappa_n^2 a^2}{\alpha^2 s^2}, \quad \beta > m_e/m_i, \tag{6.103}$$

and

$$W = -\frac{4\pi u \varkappa_n^2 a^2(\varepsilon - 1)}{(\varepsilon u^2 - \alpha^2) s^2}, \quad \beta < m_e/m_i, \tag{6.104}$$

where the s in (6.103) is given by (6.62) and in (6.104), by (6.96). From (6.102)–(6.104) it can be seen that the energy of vortex tubes can be negative in an inhomogeneous plasma ($\varkappa_n \neq 0$). In that case, if there is dissipation in the system, then the energy will decrease but grow in absolute value. From this it follows that Alfvén waves in an inhomogeneous plasma will condense into structures having the form of vortex tubes, because this is energetically more favourable.

We now consider the behaviour of vortex tubes under the influence of dissipation. The simplest kind of dissipation is the magnetic viscosity due to finite conductivity. When this effect is included, the equation for the longitudinal electron motion (6.88), valid both for finite and small β, is written as

$$\partial_t A - \varepsilon d_t J + d_z(\Phi - N) = \nu \nabla_\perp^2 A, \tag{6.105}$$

where ν is the coefficient of magnetic viscosity in dimensionless form.

In that case the energy E is no longer constant, but varies with time according to the formula

$$\partial_t E = -2\nu \int (\nabla_\perp A)^2 d^3x. \tag{6.106}$$

Assuming the dissipation to be small, one can substitute solutions in the form of the above vortices into (6.106), regarding the parameters a, u, α to be slow functions of time. It is clear from (6.106) that, because W is negative (formulas (6.102) through (6.104)), vortex tubes carry more energy as a result of dissipation (their size and rate of rotation increase). Alfvén waves are the most widespread oscillation mode in laboratory and space plasmas. They play an important part in the acceleration of particles in the Earth's magnetosphere, in turbulent mixing of plasmas, etc. When dispersion is taken into account, this mode couples with the drift mode and leads to exchange interactions between the wave and the plasma owing to inhomogeneity. As a result, the free energy of a plasma associated with an inhomogeneity is converted into vortical motion through dissipation. Finite ion Larmor radius effects can be neglected in the region where the modes intersect as they have no influence on the coupling and only the effect of the longitudinal electric field has to be included. Equations that incorporate such effects were derived above. These equations were used to show that the Alfvén waves are organized in the form of vortex tubes with exponential localization. We have shown that their energy can become negative in an inhomogeneous plasma. Therefore, vortex tube formation is energetically favourable, as is the condensation of a vapour into liquid drops. Such vortices can exist and grow in plasmas with shear which are stable in the linear approximation. They can develop from strong fluctuations. Such fluctuations can develop during RF heating, the tearing mode instability, particle injection, etc. One can, therefore, expect that accumulations of such tubes can arise in plasmas.Their size is only restricted by the magnetic shear. Because of their finite length, they can cause convective mixing of a plasma. This may be an explanation for the high heat conduction and diffusion observed in plasmas, at levels much higher than the classical values.

6.8. OBSERVATION OF ALFVÉN VORTICES IN THE MAGNETOSPHERIC PLASMA

The auroral ionosphere and the magnetosphere represent an extremely dynamic system which is controlled by interactions between energetic particle beams and the plasma.

Kinetic Alfvén waves generated by this interaction play an important role in the coupling between the ionosphere and magnetosphere.They may determine the fine structure of many auroral forms.

Wave and particle measurements with the Intercosmos-Bulgaria-1300 satellite (IC-B-1300) have shown that discrete fluxes of low energy electrons and suprathermal ions are often related to nonlinear Alfvén waves (Chmyrev *et al.*, 1988). In the following we shall describe the vortex-like electromagnetic structures detected by the IC-B-1300 satellite in the high-latitude magnetosphere that may correspond to the vortex structures described above. Six components of the electromagnetic field in the frequency range 0.1–8 Hz were measured on board this satellite (Chmyrev *et al.*, 1988). The satellite operated in a circular orbit with a typical altitude of 850 km and an inclination of 81.3°. The data were presented in orbital coordinates: x — along the spacecraft velocity vector, z — in the upward vertical direction and y — to complete the coordinate system.

Figure 6.4 presents examples of strong, small-scale magnetic field variations observed in the evening sector of the auroral zone during the magnetic storm of 2 March 1982. Here magnetic disturbances transverse to the ambient magnetic field with amplitudes up to 170 nT are shown. The electric field variations are shown on an expanded time scale in Figure 6.5. There is an apparent phase shift between the horizontal components E_x and E_y that suggests a vortex-like character of the perturbations. The variations in the horizontal components of the electric field vector along the orbit are shown in Figures 6.6 and 6.7 for the wave

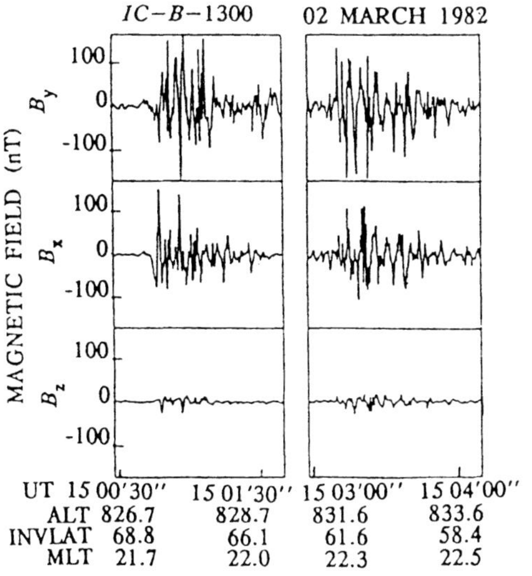

Figure 6.4. Three components of the magnetic field in the frequency range 0.1–8 Hz near the polar edge of an auroral oval (left) and in the equatorial edge of the oval (right) registered during the strong magnetic storm of 2 March 1982.

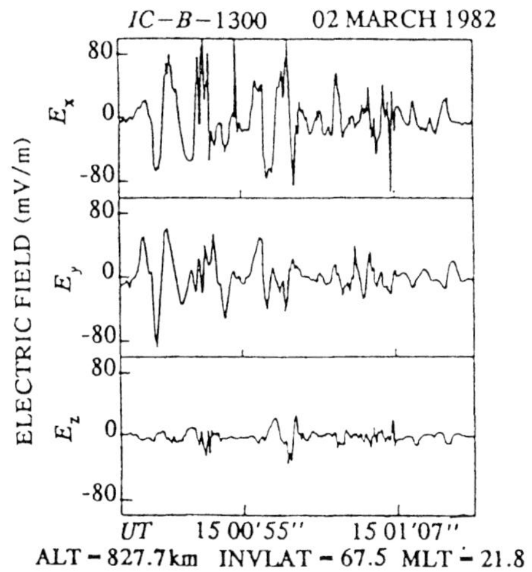

Figure 6.5. Three components of the electric field in the frequency range 0.1–8 Hz in an expanded time scale for the middle of the first interval in Figure 6.4.

trains presented in Figure 6.4. The distributions of the field vector look like vortex chains in both cases. Left- and right-hand polarization patterns can be observed in successive cells of these chains. A regular counterclockwise rotation of the electric vector is seen in the cells with a

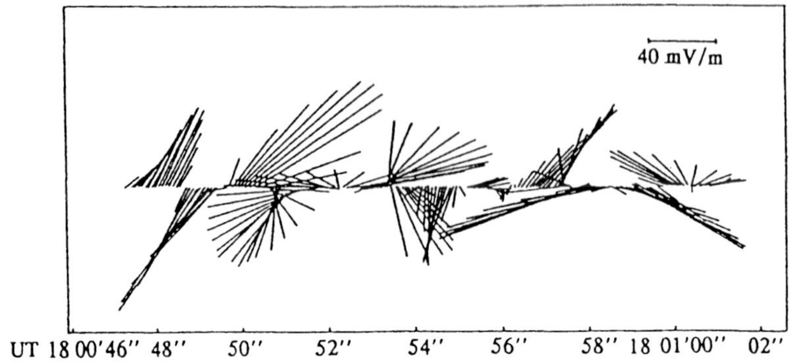

Figure 6.6. The variation of the horizontal component of the electric field vector in the frequency range 0.12–0.8 Hz along the satellite orbit in the first interval of Figure 6.4.

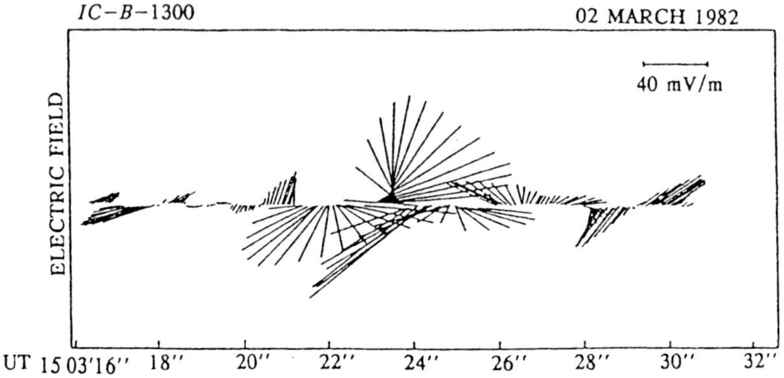

Figure 6.7. Variations of the horizontal component of the electric field vector in the frequency range 0.12–0.8 Hz along the satellite orbit in the second interval in Figure 6.4.

strong field lasting from the 21st to the 26th second (Figure 6.7). The rotation then changes to the opposite sense. A similar periodic transformation from clockwise to counterclockwise polarization is observed in Figure 6.6. Solitary vortices of both dipolar and monopolar types were observed. They are presented as hodograms in Figures 6.7 and 6.9. The polarization reversal displayed in Figure 6.8 corresponds to a two-cell structure with opposite plasma rotation within the cells. It is natural to identify it with a solitary dipole vortex. The second vortex in Figure 6.9 does not change polarization and may be regarded as a monopolar vortex. These two-dimensional localized electromagnetic

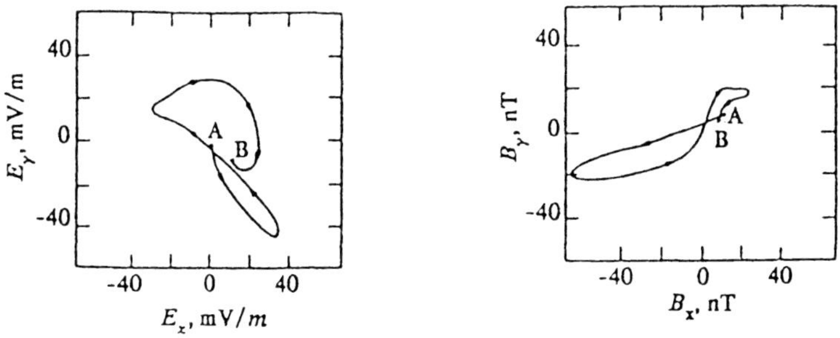

Figure 6.8. A hodograph of the low-frequency ($f \sim 0.7$ Hz) spectral component of the electric (left) and magnetic (right) fields in the interval 19 08′11″ – 19 08′14″ UT.

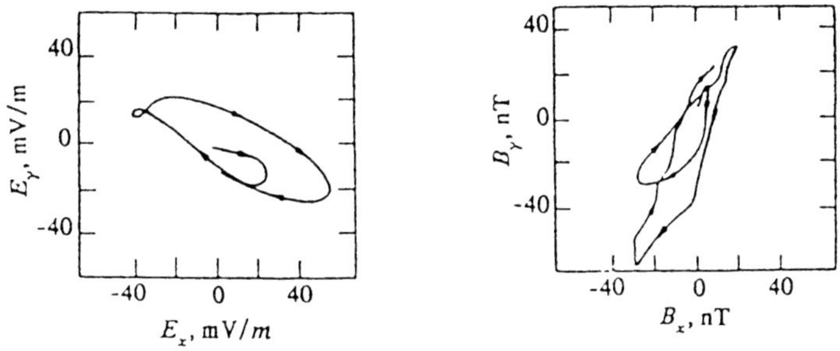

Figure 6.9. A hodograph of the low-frequency ($f \sim 0.7$ Hz) spectral component of the electric (left) and magnetic (right) fields in the interval 19 08′29″ – 19 08′32″ UT.

structures and the vortex chains shown in Figures 6.6, 6.7 are typical of the regular large-amplitude electromagnetic disturbances observed with the IC-B-1300 satellite in the auroral magnetosphere. Since $\delta B_z \ll \delta B_\perp$ and $\delta E/\delta B_\perp \approx C_A/C$, these waves are of the Alfvén type. According to Chmyrev *et al.* (1988) the propagation velocity of these vortices is approximately 15–30 km/s, which is less than the ion thermal velocity and greater than the velocity of the satellite. They may correspond to the structures described in the previous sections.

6.9. DISSIPATIVE GENERATION OF ELECTRON DRIFT VORTICES

The modern theory of anomalous transport in plasmas (Kadomtsev and Pogutse, 1984) predicts that the main contribution to the electron thermal conductivity will be from suprathermal fluctuations with a scale length of the order of the skin depth. This occurs because the electrons are no longer frozen into the magnetic field over these lengths. In the linear approximation, however, perturbations of the magnetic field of this size are stable. Petviashvili and Pogutse (1986) have shown that nonlinear effects can give rise to growing solitary structures in the form of vortex tubes which differ from the solitary Alfvén vortices considered above in having a small diameter (much smaller than r_{Bi}). It turns out that these vortices travel at a velocity below the drift velocity. Therefore, their amplitude can grow owing to Landau damping or electron collisional dissipation. This phenomenon is similar to the linear drift-dissipative instability of electrostatic drift waves (see chapter 1). These waves grow because their propagation velocity is less than the drift velocity in the linear case.

Since the size of these vortices is less than r_{Bi} and their frequency is small compared to ω_{Bi}, the ion density n in the electric potential of a vortex, φ, has the Boltzmann distribution

$$n = n_0(1 + \varkappa_n x - \tau\Phi), \quad \tau = T_e/T_i, \qquad n_0, \varkappa_n = \text{const.} \tag{6.107}$$

The electron component can, according to (6.48), (6.49), be described by the following system of dimensionless equations:

$$d_t N = -d_z \nabla_\perp^2 A, \tag{6.108}$$

$$-(1 + \nu)E_\parallel = d_z N, \tag{6.109}$$

where (6.108) is an equation of continuity for electrons with the longitudinal current taken into account and (6.109) is the equation of motion for electrons in the magnetic field which reduces to the balance between the gradient of electron pressure and the longitudinal electric field $E_\parallel = -d_z\Phi - \partial_t A$ in the present case, where the wave velocity along z is small compared to v_{Te}. As in the preceding, we change to dimensionless variables in (6.108) and (6.109). The operator ν in (6.109) described dissipation on the electrons in the linear approximation. A comparison with the linear theory (see § 1.3) readily yields the Fourier spectrum of this operator:

$$\nu_{k,\omega} = i\sqrt{\pi}(\omega + \varkappa_n k_y)/|k_z| v_{Te} \tag{6.110}$$

Using the quasi-neutrality condition, we substitute (6.107) into (6.108) and (6.109). Disregarding dissipation for the moment, we obtain a set of equations for small-scale drift waves:

$$\tau\partial_t\Phi + \varkappa_n\partial_y\Phi = d_z\nabla_\perp^2 A, \tag{6.111}$$

$$\partial_t A - \varkappa_n\partial_y A = -(1+\tau)d_z\Phi. \tag{6.112}$$

The dispersion relation (1.39) follows from this in the linear approximation. It describes two oscillation modes (branches) travelling in the direction of the ion or electron drift velocity, respectively. Both branches are stable, because their velocities exceed the drift velocities. However, as we shall see later, including the nonlinearity makes these branches coalesce to form vortices whose velocity is between the ion and electron drift velocity. As a consequence, they can grow as a result of Landau damping. This can be seen from the expression for the Fourier component of the damping operator (6.110). It changes sign when $|\omega| < |\varkappa_n k_y|$. We now find a steady-state 2D solution to (6.111) and (6.112) that travels obliquely to the magnetic field at velocity u. Suppose all quantities are functions of x and $\eta = y + \alpha z - ut$. Then, from (6.112) we get

$$\Phi(x,\eta) = \frac{u + \varkappa_n}{\alpha(1+\tau)} A(x,\eta). \tag{6.113}$$

The constant of integration in (6.113) is determined from the requirement that Φ and A decrease at infinity. Using (6.113), we obtain from (6.111):

$$\nabla_\perp^2 A + \alpha s^2 x = f(A - \alpha x),$$
$$s^2 \equiv \frac{(\varkappa_n + u)(\varkappa_n - \tau u)}{\alpha^2(1+\tau)}, \tag{6.114}$$

where f is an arbitrary function. Equation (6.114) is solved by the Larichev–Reznik method. For the equation to have a sufficiently localized solution we must have $s^2 \gg 1$, which happens if the propagation velocity lies in the interval between the ion $(\varkappa_n/\tau)$ and electron drift velocities $(-\varkappa_n)$. We define the coordinate $r = (x^2 + \eta^2)^{1/2}$ and represent f as a linear function with different coefficients inside and outside the circle of radius $a \ll r_{Bi}$. The coefficients are chosen so that a localized solution of (6.114) of the form

$$A = b_0 F_0(r) + \alpha a F_1(r) x/r \tag{6.115}$$

is obtained. Here F_0, F_1 are continuous along with their first derivatives; they can be expressed in terms of Bessel functions inside the circle and Macdonald functions outside it. One should set $A - \alpha x =$ constant at the matching boundary $r=a$. The amplitude b_0, angle α, radius a, and velocity u remain arbitrary. The solution (6.115) falls off at infinity as exp $(-sr)$.

We now investigate the effect of dissipation on this solution. To do this, we note that the system (6.111) and (6.112) conserves the integral

$$W = \int [(\nabla_\perp A)^2 + \tau(1+\tau)\Phi^2] d^3x. \tag{6.116}$$

If dissipation is taken into account as in (6.109), then this integral varies with time:

$$\partial_t W = -\int \nabla_\perp^2 A \nu E_\parallel \, d^3x. \tag{6.117}$$

Assuming that the dissipation is small, then we can substitute the solutions (6.113) and (6.115) in (6.117). Writing (6.117) in dimensionless form and using the fact that $\omega = k_y u$ in the steady-state case we find that

$$\partial_t W = |\alpha| s^2 c_A v_{Te}^{-1} \int k^2 |k_y| \, |A_k|^2 d^3k, \tag{6.118}$$

for a steady-state vortex in accordance with (6.113). We see from this that W increases with time. The amplitude b_0 can grow with a, u, α unchanged. A plateau forms as the amplitude increases, and Landau damping transforms to weak collisional dissipation (Artsimovich and Sagdeev, 1979). This retards vortex growth somewhat.

It is thus clear that when the nonlinearity is included, packets of electron drift waves of size smaller than r_{Bi} form vortex tubes (electron vortices) which propagate more slowly than the drift velocity. Because of this, electron dissipation causes these vortices to grow until their energy density becomes comparable to the thermal energy. Kadomtsev and Pogutse (1984) have obtained transport coefficients for such fluctuations.

6.10. FLUTE VORTICES

The preceding sections have been concerned with vortices whose longitudinal velocities were on the order of the Alfvén velocity. The dispersion and nonlinear properties of these waves permit the existence of self-localized 2D wave packets whose size is bounded from below only by the ion Larmor radius or the skin depth. This strong localization makes it possible to neglect inhomogeneities of the ambient magnetic field.

This localization mechanism does not play a significant role in flute perturbations, in which the velocities of propagation along $\mathbf{B}_0$ are large compared with v_{Te} and c_A. It then turns out that localization is ensured by the curvature of the field lines of the ambient magnetic field (Pavlenko and Petviashvili, 1983; Ivanov and Pokhotelov, 1988). Effects due to the shear play a significant role in toroidal plasma devices along with the curvature.

There is a difficulty in describing oscillations of inhomogeneous plasma in a curved magnetic field in the linear approximation, because the quasi-classical approximation is not valid. However, if the wave amplitude is finite, it turns out that wave packets localize themselves. This simplifies the problem, since one can then use a local approximation with the small parameter given by the ratio of the packet size to the inhomogeneity scale length. Flute vortices are nearly parallel to the magnetic field lines in the unperturbed field. For this reason, we shall describe them by using equations (6.47)–(6.49) with $\partial_z = 0$. For simplicity the curvature of the magnetic field is taken into account by defining an effective acceleration g in the vorticity equation (6.47). The latter is then rewritten as

$$d_t \nabla_\perp^2 \Phi + \operatorname{div}\{P_i, \nabla_\perp \Phi\} + g\partial_y P = \{A, J\}, \qquad (6.119)$$

where $P = P_i + P_e$ is the total plasma pressure. As for the other equations, we just set $\partial_z = 0$ in them. Then we have

$$d_t P_e = \{A, J\}, \quad d_t P_i = 0, \quad d_t A = \{P_e, A\}. \qquad (6.120)$$

In the linear approximation the system (6.119) and (6.120) leads to the dispersion relation for flute oscillations (1.40). This system has the energy integral

$$E = \int [(\nabla_\perp \Phi)^2 + (\nabla_\perp A)^2 + P_e^2 - 2\varkappa g P + 2\varkappa^2 g(\varkappa_e + \varkappa_i) - \varkappa_e^2 x^2]d^2x. \qquad (6.121)$$

The following integrals are also conserved:

$$\int f(A)d^2x, \quad \int AP_e d^2x, \quad \int f(P_i)d^2x, \quad \int P_e d^2x. \qquad (6.122)$$

Here we have used the fact that $P_{e,i} \to \varkappa_{e,i}\, x$ far from the disturbances.

We shall seek a localized solution of (6.119) and (6.120). We assume that it travels along y at velocity u. Under the assumption that the vortex is local, u turns out to be an eigenvalue of the problem, and the solution is

$$P_i = \varkappa_i(x - \Phi/u), \quad A = A(x), \qquad (6.123)$$

and

$$P_e = \varkappa_e(x - \Phi/u), \quad u = u_e = -\varkappa_e. \qquad (6.124)$$

Note that the solutions (6.123) and (6.124) exist only for $u = u_e$, where u_e is the dimensionless electron drift velocity. From this one can see that when a wave travels at the electron drift velocity, A is an arbitrary function of x, and the electric field is an electrostatic field. This means that vortices travelling at velocity u_e are not sensitive to arbitrary strong shear. When $u = u_e$, the magnetic field does not depend on time. Since $B_y(x)/B_0 = -\partial_x A$, $B_y(0) = 0$ and $B_x = 0$, the magnetic field is not perturbed, because it is frozen into the electrons which drift with the same velocity. Substituting (6.123) and (6.124) into (6.119) and integrating, we obtain

$$\nabla_\perp^2 \Phi = xg + F(\Phi - xu_e), \qquad (6.125)$$

where F is an arbitrary function. This equation is solved by the Larichev–Reznik method and the solution is of the form (5.43). Here, one should set $\varkappa^2 = g/u_e$ in (5.41) and (5.42). Because of this, flute vortices under shear have one free parameter less, since their velocity is an eigenvalue of the problem. Since a vortex travels at the electron drift velocity u_e, it follows that the electric field is independent of the time in a reference frame travelling at this velocity, so that there is no resonant interaction with the electrons. Note that a vortex falls off exponentially with distance from the centre. As the diameter a increases, so does the effective wavenumber along $\mathbf{B}_0$, until ion Landau damping becomes operative. The phase velocity of a vortex is of order $u_e/(k_\parallel a)$ along the magnetic field and must be greater than the ion thermal speed. Given that $k_{\parallel} \simeq k_{\perp} a/L_s$ at the edge of a vortex (L_s is the typical shear length) and that $k_{\perp} a \simeq 1$ in our case, we obtain an estimate of $a < r_s L_s/L_p$ for the maximum vortex size. Note that a vortex can exist only where $g < 0$, that is, in the region with an unfavourable magnetic drift, where the gradient of the magnetic field points in the same direction as the gradient of the plasma pressure. The axis of a vortex is parallel to the magnetic field in the central part of the vortex, while at the periphery, where the magnetic field is inclined because of the shear, the vortex corresponds to the Alfvén mode. For this reason it is natural to refer to such vortices as flute-Alfvén vortices. In the absence of shear potential flute vortices move at a velocity higher than the ion drift velocity if $\varkappa_i g < 0$ (unstable plasma) and slower than that in the opposite case.

Here we show that slow vortices ($\varkappa_i g > 0$) may be amplified through the ion viscosity. Thus, even in the stability region of a plasma, flute waves can contribute substantially to anomalous transport of particles through vortex convection. Suppose for simplicity that $p_i = p_e = \varkappa_i x + p$. Then a system of model equations for the relative perturbations of the pressure p_i and the electric potential Φ follows readily from (6.119) and (6.120) if we take $A = 0$:

$$d_t \nabla_{\perp}^2 \Phi + \operatorname{div}\{p, \nabla_{\perp} \Phi\} + 2g\partial_y p = \mu \nabla_{\perp}^4 (\Phi + p), \qquad (6.126)$$

and

$$\partial_t p - \varkappa_i \partial_y \Phi = \{\hat{}p, \Phi\}. \qquad (6.127)$$

Here we have introduced an ion viscosity term on the right-hand side of (6.126) in the form suggested by Mikhailovskii (1979). The viscosity effect is proportional to the deviation from a Boltzmann distribution. In the absence of viscosity ($\mu = 0$), the system (6.126) and (6.127) has the energy integral

$$E = \int \left\{ (\nabla_{\perp} \Phi)^2 + 2 \frac{g}{\varkappa_i} p^2 \right\} d^2 x \, . \qquad (6.128)$$

On including the viscosity, we obtain

$$\partial_t E = -2\mu \int \int [(\nabla_{\perp}^2 \Phi)^2 + \nabla_{\perp}^2 \Phi \nabla_{\perp}^2 p] d^2 x. \qquad (6.129)$$

In the linear approximation, for $\varkappa_i\, g < 0$ the equations (6.126) and (6.127) describe the flute instability. When the nonlinearity is included, they have solutions in the form of solitary flute vortices. If a vortex travels at velocity u along y, then (6.127) implies that

$$p = -(\varkappa_i / u)\Phi, \quad \Phi = \Phi(x, y - ut)\,. \tag{6.130}$$

Substituting (6.130) into (6.126), we obtain

$$\nabla_\perp^2 \Phi - \varkappa^2 ux = F(\Phi - ux), \tag{6.131}$$

where $\varkappa^2 = 2g\varkappa_i / u(u - \varkappa_i)$ and F is an arbitrary function. Solitary solutions exist for $\varkappa^2 > 0$ only. Then this equation has a solution of the Larichev–Reznik type.

When dissipation is included the energy of the solution changes in accordance with (6.129). Substituting (6.130) in (6.129), we see that in a stable plasma, where $\varkappa_i / u > 1$, the right-hand side of (6.129),

$$\partial_t E = -2\mu \left(1 - \frac{\varkappa_i}{u}\right) \int (\nabla_\perp^2 \Phi)^2 d^2x,$$

is positive. This expression suggests that in a stable plasma, flute vortices can be amplified through the ion viscosity. This idea has been tested numerically. Numerical solution of the system (6.126) and (6.127) shows that in the presence of flute instability ($\varkappa_i\, g < 0$), the vortices are unstable even in the absence of viscosity, and they break up over a time on the order of the time for development of the flute instability. A calculation for the stable plasma case shows that when $\mu = 0$, solitary vortices of Larichev–Reznik type do not decay; in fact, they retain their shape. For $\mu > 0$ the vortices began to grow. Since the pressure is frozen in the plasma according to (6.127), we should expect an increase in $|\nabla_\perp \Phi|$, i.e., an increase in the plasma rotation velocity in a vortex. The displacement velocity u and the shape of the vortex should vary slowly.

The numerical calculations show that the maximum vortex potential does indeed increase at first and then reaches saturation. The maximum density falls off slightly in the process (Figure 6.10). The vortex energy (6.128), however, continues to grow, because of the increase in the potential gradient, and thus, the particle rotation velocity in the vortex. The vortex remains topologically stable, although it deviates significantly

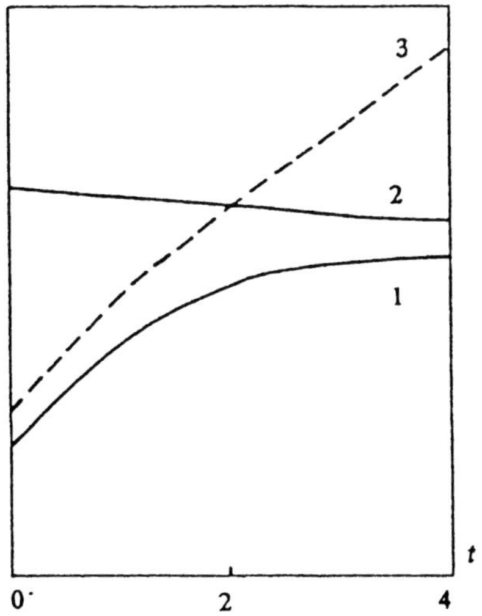

Figure 6.10. The time evolution of the maximum value of the potential (curve 2) and of the energy integral (curve 3) in a fluid vortex owing to viscosity. The structure of the vortex deviates significantly from the initial Larichev–Reznik solution where the velocity of the vortex is half of the ion drift velocity.

from the steady-state solution for the case without viscosity.

6.11. SOLITARY BALLOONING VORTICES

We have discussed flute vortices without a longitudinal current in which the magnetic field is not perturbed. When perturbations of the longitudinal current are included, they become balloon vortices. Such vortices require a more accurate accounting for the geometry of the magnetic field. As in § 1.4 where the flute instability was analysed, we shall investigate balloon vortices by using Euler's coordinates for a curved magnetic field. A solution in these coordinates (Ivanov and Pokhotelov, 1988) allows one, first, to describe the real spatial structure of the vortices (which is different from that in a straight field line geometry) and, second, to include the effects of the plasma compressibility more accurately. This already shows up in the linear approximation (see § 1.4).

Our study of ballooning modes is based on the closure equation for the current (6.42), the equation for the longitudinal motion of the electron without inertia (6.45), and the equation for the ion and electron pressure which, in contrast to Alfvén waves, include the effect of the compressibility drift velocity v_E associated with the curvature of the field. When the drift effects are neglected, the last two equations take the form

$$d_t p_i + \gamma p_i \operatorname{div} \mathbf{v}_E = 0,$$

$$d_t p_e + \gamma p_e \operatorname{div} \mathbf{v}_E - \gamma p_e \operatorname{div}\left(g\mathbf{j}_{\parallel}/en + \frac{\mathbf{B}_{\perp}(\mathbf{j}_{\parallel}\cdot\mathbf{B}_0)}{B_0^2} \right) = 0, \tag{6.132}$$

where the oscillations of longitudinal current have been included. The unperturbed curved field $\mathbf{B}_0$ is described using Euler's coordinates x_0^1, x_0^2, x_0^3 by a relation of the form (1.54). We assume them to be mutually orthogonal, that is, $g_{ik} = 0$ when $i \neq k$. It is then convenient to transform to the oblique coordinates x^1, x^2, x^3 according to the transformation

$$x^1 = x_0^1, \quad x^2 = x_0^2 + \alpha x_0^3, \quad x^3 = x_0^3, \tag{6.133}$$

where α is a constant. The magnetic field B_0 in the new coordinates has a component $B_0^2 = \alpha/\sqrt{g}$ along the x^2-axis. The electric and transverse magnetic field are described with the help of the potentials φ and A.

$$\mathbf{E} = -\nabla\Phi - c^{-1}\partial_t A \nabla x^3, \quad \mathbf{B}_{\perp} = \nabla A \times \nabla x^3, \tag{6.134}$$

Integrating the original equations over x^3 as in § 1.4, we arrive at the following set of equations (Ivanov and Pokhotelov, 1988):

$$\{V, p_e + p_i\} + \rho_0\, c\, d_t \Delta_2 \varphi + (\alpha/4\pi) d_{\parallel} \Delta_2' A = 0, \tag{6.135}$$

$$\partial_t A + \partial_1 p_{0\,e} \partial_2 A + c\,\alpha\, \partial_{\parallel} (\varphi - p_e/en_0) = 0, \tag{6.136}$$

$$V \partial_t p_{Ve,i} + c\gamma p_{0\,e,i}\{\varphi, V\} = 0, \tag{6.137}$$

$$d_t p_{pe,i} + c\{\varphi, p_{0\,e,i}\} = 0, \tag{6.138}$$

and

$$d_t p_{de} = c\,\gamma\, p_{0\,e}(\alpha/4\,\pi e n_0) d_{\parallel} \Delta_2' A = 0. \tag{6.139}$$

The notation is the same as in § 1.4. We also define

$$d_{\|} \equiv \partial_z - \alpha^{-1}\{A, \dots\} \quad \text{and} \quad \Delta_2' \equiv V_{11}'\partial_2^2 + V_{22}'\partial_1^2.$$

The perturbed ion and electron pressures are represented in (6.135)–(6.138) in the form $p = p_v + p_p + p_d$, that is, as sum of the pressure perturbations due to the inhomogeneities in the magnetic field and pressure, respectively, while the last term is related to dispersion effects. In contrast to the case of straight field, the effects of plasma compressibility are significant in a curved field, div $\mathbf{v}_E \neq 0$, and give rise to an extra term in the perturbed pressure, p_v. The quantities V, V_{ik}, V_{ik}' are proportional to the longitudinal perturbation lengths, so that when the transverse dimensions are small compared with the longitudinal ones in equations (6.135)–(6.139), one can neglect plasma edge effects, in contrast to § 1.4. We shall also assume that the perturbations are on a small scale compared with the typical dimensions of plasma pressure and field inhomogeneities.

For definiteness we assume that $p_{0e,i}$ and V are functions of x^1 only. In that case (6.135)–(6.139) have the energy integral:

$$W = \int \Big\{ \rho_0 c^2 \Big[V_{11}(\partial_2\varphi)^2 + V_{22}(\partial_1\varphi)^2 + (1/4\pi)\, [V_{11}'(\partial_2 A)^2 + V_{22}'(\partial_q A)\Big] + \sum_{\alpha=e,i} \left(\frac{V}{\gamma} \frac{(p_{v\,\alpha})^2}{p_{0\,\alpha}} + \frac{(\partial_1 V)(p_{p\,\alpha})}{\partial_1 p_{0\,\alpha}} \right) + \frac{V}{\gamma p_{0e}} [p_e^2 - (p_e - p_{de})^2]\; dx^1 dx^2. \tag{6.140}$$

The last term in (6.140) is controlled by the dispersion part of the electron pressure perturbation p_{de}, which will be disregarded in the following. It has been included in order to permit an accurate calculation of the perturbation energy. We shall show that (6.135)–(6.139) have solitary solutions that travel at velocity u along x^2. The perturbed quantities are assumed to be functions of x^1 and $\eta = x^2 - ut$. We then obtain the following from (6.135)–(6.139) in the steady-state case:

$$\rho_0 c d_2 \Delta_2 \varphi - (\alpha/4\pi) d_{11} \Delta_2^1 A - \frac{\partial_1 V \partial_\eta (p_e + p_i)}{u} = 0, \tag{6.141}$$

$$d_{11}\left[(1-u^{-1}\partial_1 p_{0e})A - \alpha c u^{-1}(\varphi - p_e/en_0)\right] = 0, \tag{6.142}$$

$$d_2\left[p_{Ve,i} + c\,\gamma p_{0e,i}\frac{\partial_1 V}{uV}\varphi\right] = 0, \tag{6.143}$$

$$d_2\left[p_{pe,i} + c\,\frac{\partial_1 p_{0e,i}}{u}\varphi\right] = 0, \tag{6.144}$$

where $d_z \equiv \partial_\eta - (c/u)\{\varphi, \dots\}$.

Equations (6.141) through (6.144) are satisfied under the following conditions:

$$p_{Ve,i} = -\frac{c\,\gamma p_{\,0e,i}}{u}\frac{\partial_1 V}{V}\varphi,$$

$$A = \frac{\alpha c}{u}\varphi, \tag{6.145}$$

$$p_{pe,i} = c\,\frac{\partial_1 p_{0\,e,i}}{u}\varphi.$$

Substituting (6.145) into (6.141), we arrive at the equation

$$\partial_z^2(\varphi - \lambda^2\Delta_\perp\varphi) + \frac{c\lambda^2}{u l_1^{1/2}}\{\varphi, \Delta_\perp\varphi\} = 0, \tag{6.146}$$

where $\lambda^2 = u^2/\Gamma^2$, Γ is given by (1.61) and $\Delta_\perp \equiv \partial_{z^1}^2 + \partial_{z^2}^2$. The following similarity transformation has been made in (6.146):

$$x^1 = \left[V_{22}\left(1 - \frac{\alpha^2 c_{A2}^2}{u^2}\right)\right]^{1/2} z^1 = l_1^{1/2} z^1,$$

$$x^2 = \left[V_{11}\left(1 - \frac{\alpha^2 c_{A1}^2}{u^2}\right)\right]^{1/2} z^2 = l_2^{1/2} z^2. \tag{6.147}$$

The quantities $c_{A1} \equiv (V_{11}/4\pi\rho_0 V_{11})^{1/2}$ and $c_{A2} = (V^{\,1}_{\,22}/4\pi\rho_0 V_{22})^{1/2}$ are ana-logous to the Alfvén velocity in a straight magnetic field. Equation (6.146) has a Larichev–Reznik solution as a superposition of a carrier and a rider of arbitrary amplitude (see § 5.6). The shielding condition $\lambda^2 > 0$

must also be satisfied. This condition coincides with the criterion for the flute instability (1.41) $\Gamma^2 > 0$.

The similarity transformation (6.147) implied that the vortex velocity satisfied the condition $u^2 > \max(\alpha^2 c_{A1}^2, \alpha^2 c_{A2}^2)$, that is, $l_1 > 0$, $l_2 > 0$. However, the system (6.135)–(6.139) also has solitary solutions in the case $u_2 < \min(\alpha^2 c_{A1}^2, \alpha^2 c_{A2}^2)$. To demonstrate this, we make the transformation $x^1 = (-l_1)^{1/2} z^1$, $x^2 = (-l_2)^{1/2} z^2$ instead of (6.147). The system (6.135)–(6.139) then reduces to a nonlinear equation similar to (6.146) with the difference that the coefficient of λ^2 has the opposite sign. For a solitary solution to exist in this case, we must also have $\Gamma^2 < 0$.

Taking (6.145) into account, we can write the expression for the energy of vortical tubes (6.140) in the form

$$W = \rho_0 c^2 \int \left[V_{11} \left(1 + \frac{\alpha^2 c_{A1}^2}{u^2} \right) (\partial_2 \varphi)^2 \right.$$
$$\left. + V_{22} \left(1 + \frac{\alpha^2 c_{A2}^2}{u^2} \right) (\partial_1 \varphi)^2 - \lambda^{-2} \varphi^2 \right] dx^1 dx^2. \tag{6.148}$$

One can see that for $\Gamma^2 > 0$, that is, or flute instability, the energy of the vortical tubes can become negative. Only sufficiently large vortices can possess this property. Their typical dimensions must exceed $d_{cr} = \max\{[V_{11}(u^2 + \alpha^2 c_{A1}^2)/\Gamma^2]^{1/2}, \quad [V_{22}(u^2 + \quad + \alpha^2 c_{A2}^2)/\Gamma^2]^{1/2}\}$. The effects described above can take place in the magnetospheric plasma (Ivanov and Pokhotelov, 1988). The latter is generally stable with respect to the generation of flute and ballooning modes. However, powerful nonstationary processes leading to large pressure gradients can create conditions under which vortical tubes will be spontaneously generated. We note that the effects of plasma compressibility in a curved magnetic field play an important role in these processes. If these are not included, then the plasma will always be unstable in machines with a magnetic field that decreases with distance from the centre. For this reason, modelling the curvature by using gravity does not adequately describe the behaviour of the plasma in the magnetosphere.

6.12. VORTEX CONVECTION AND ANOMALOUS RESISTIVITY IN PLASMAS

When a plasma is unstable with respect to the excitation of electrostatic drift and flute-Alfvén waves, the entire unstable region seems to be filled with the vortices considered in the preceding sections. They correlate weakly with one another owing to self-localization, so they are distributed randomly. Vortical turbulence of this sort can be seen clearly in rapidly rotating shallow water (Figure 5.9). Here one can see the particles in a random walk between vortices. Trapped particles are rotating in the central part of a vortex and also contribute to convection. The vortices themselves move along random paths during the interaction.

In the linear approximation, a plasma can be stable with respect to drift wave excitation owing to shear stabilization, finite ion Larmor radius effects etc. However, if strong perturbations operate in such plasma for some time, for example, injection of neutrals, RF heating, etc., then vortices of the type considered above may remain after the perturbations have been switched off. They are not sensitive to shear and other stabilizing factors and can be maintained by the dissipative instability. It is known, for example, that increased thermal conductivity and diffusion are observed in plasmas after intensive RF heating. Nekrasov, Pavlenko and Petviashvili (1985) have shown that a heating electromagnetic wave is unstable with respect to decay into an electromagnetic and a drift Alfvén wave. The rate of wave decay can overcome the influence of the stabilizing factors mentioned above. As a result of the action of the nonlinearity, 2D flute-drift vortices may form from the excited drift Alfvén waves. According to Nekrasov, Pavlenko and Petviashvili (1985), the growth rate of the decay instability is of order

$$\gamma \simeq \frac{v_0}{r_s(kL_p)^{1/2}}, \quad v_0 = \frac{eE_0}{m_e\omega_0}, \tag{6.149}$$

where L_p is the typical scale length of the plasma inhomogeneity and v_0 the velocity of the electron oscillations in the field of the heating wave. At available heating power levels, the growth rate (6.149) exceeds the drift frequency, so that the instability may be aperiodic in character and thereby weaken the stabilizing effect of shear. As a result, flute-Alfvén vortices develop from the excited drift Alfvén waves as their amplitude increases. The transverse dimension of the vortices is of order $a \sim r_s L_s / L_p$, where L_s is the typical shear scale length (large vortices are attenuated by the ion Landau damping mechanism), while the velocity at which they propagate across B_0 is equal to the electron Larmor drift velocity. The typical rate of plasma rotation in a vortex is also of the order u_e. For mixing of untrapped particles it is significant that the vortices are not parallel to one another owing to the shear. The typical mixing scale length in a dense packing of such vortices is of order L^2/L_s, where L is the typical packing scale length (approximately equal to the dimensions of the inhomogeneous region). Hence the convective heat conductivity is given roughly by

$$x \sim u_e L^2/L_s = D_B L^2/L_p L_s, \tag{6.150}$$

where $D_B = (v_{Ti} r_s)$ is the Bohm diffusion coefficient. The scale length L remains indeterminate in the theory. One can only say that $L \lesssim L_p$. For this reason the convective thermal conductivity is always smaller than the Bohm value, because $L_s \gg L_p$. The increased thermal conductivity observed in experiments can probably be explained by clusters of these vortices, which lead to convective mixing of plasma with a large characteristic size.

It has been observed in experiments with discharges that when the current velocity exceeds some threshold, an additional electric resistivity appears in the plasma. This is the so-called anomalous resistivity. It can be large compared to the classical value due to ordinary Coulomb collisions. A current can be transported in a plasma owing to the acceleration of electrons in the tail of the distribution function whose collision rate is small compared to the thermal electron collision rate. This process can generate runaway electrons that do not experience Coulomb collisions (Stefanovsky, 1965). Budker (1956) has suggested using this effect to produce intense relativistic beams in a toroidal plasma. This type of experiment was conducted in many laboratories all over the world during the late 1950s. These installations were referred to as plasma betatrons. The first series of experiments, however, revealed that intense beams of runaway electrons do not form, even in very strong electric fields. Powerful oscillations develop in the plasma instead. Kadomtsev and Pogutse (1968) explained this phenomenon by the appearance of additional friction owing to scattering of runaway electrons caused by the generation of lower hybrid oscillations.

Another mechanism for current formation is movement of the electron gas as a whole relative to the ions. Here, also, on reaching a current velocity u of order v_{Ti}, an anomalous resistivity arises that prevents further growth of u. Petviashvili (1964) and Kadomtsev (1965) have explained this effect in terms of the excitation of ion acoustic oscillations by an electron current. This mechanism is only feasible, as is generally known, in highly nonisothermal plasma where $T_e \gg T_i$. However, anomalous resistivity is also frequently observed when $T_e \simeq T_i$, when ion acoustic oscillations cannot be excited because of strong ion Landau damping. This phenomenon was explained by Drummond and Rosenbluth (1962) as the excitation of the first harmonic of electrostatic ion cyclotron waves, for which $k_\perp \gg k_\parallel$ by the current. However, there is some indirect evidence that the deceleration of the electron by the ion cyclotron waves is not efficient enough: the absence of a well-defined maximum near the ion cyclotron frequency in the turbulence spectrum observed in plasmas with anomalous resistivity. Observations show that this spectrum includes nearly equally strong frequencies below the ion cyclotron frequency. In the following, we consider the possibility that solitary vortices with frequencies $\omega \ll \omega_{Bi}$ develop in the nonlinear stage of this instability because energy is pumped into the low frequency range into both the Alfvén and the ion acoustic branches. Vortices in the ion acoustic branch (electrostatic vortices) can also exist for $T_e \simeq T_i$, unlike ion acoustic waves, because the velocity of vortex propagation along the magnetic field can be greater than v_{Ti}. The range of longitudinal velocities for both types of vortices is rather broad and extends up to v_{Te}. For this reason, they can be strongly amplified by the electric current in regions where the derivative of the electron velocity distribution function is positive. There is a fairly strong electric field along the magnetic field (along z) in kinetic Alfvén, and electrostatic vortices (§ 6.6). Therefore, they include particles trapped in the z-direction (both ions and electrons). Friction between untrapped particles and the trapped ones in the longitudinal electric field is the principal mechanism for the anomalous resistivity. As the vortices are distributed randomly, electrons moving along the z axis are decelerated in the same manner as in a random electric field. They can, therefore, be described using the quasi-linear approximation employed by Galeev and Sagdeev (1979) to evaluate anomalous resistivity in the ion acoustic turbulence. The only difference is that Galeev and Sagdeev (1979) assumed that the random longitudinal electric field is caused by ion acoustic turbulence, while in an isothermal plasma it is caused by the appearance of vortical structures. Note that vortices excited by the electric current can travel across the magnetic field much faster than the drift velocity. For this reason, isothermal plasmas can be assumed to be homogeneous when dealing with anomalous resistivity in them and we can assume that drift effects are negligible ($u_e = 0$).

7. Solitary toroidal vortices

7.1. SOLITARY STEADY-STATE VORTICES IN IDEAL HYDRODYNAMICS

So far we have dealt with solitary structures in which the shielding effect was due to dispersion. A high degree of localization (exponential fall-off at infinity) was attained thereby. Media without dispersion but with a large enough number of integrals of motion can also have solitary solutions whose constant shape is related to the existence of these integrals. These solutions are called topological solitons. We shall be interested in topological solitons in ordinary hydrodynamics and magnetohydrodynamics. We note that travelling topological solitons are localized according to a power law, so that they interact at a distance. At the same time, solitons at rest may be localized exponentially or more strongly.

There is a well-known method based on a statistical description of solutions to the equations of hydrodynamics. It is based on the hypothesis that nonlinearity in incompressible fluids produces complete randomization of all the degree of freedom. Substantially simplifying the equations of hydrodynamics, Lorentz reduced them to just three ordinary differential equations involving three independent parameters (Rabinovich and Trubetskov, 1984), which describe the nonlinear stage in the development of the convective instability. When the parameters have values in some bounded region, solutions of the Lorentz equations become completely stochastic (are randomized). In that case, any point moving in phase space describes a three-dimensional curve that tends toward some finite region of phase space known as a stochastic or strange attractor. Phase paths no longer leave the attractor. Extending this picture to the case of a nondenumerable number of degrees of freedom (the continuum case) is not straightforward. A more complex situation may possibly occur in hydrodynamics: some degrees of freedom correlate rigidly with one another as 3D topological solitons, while the latter are distributed randomly.

We begin our investigation of topological solitons by examining Euler's equations:

$$d_t \mathbf{v} = -\nabla p/\rho, \tag{7.1}$$

$$\text{div } \mathbf{v} = 0. \tag{7.2}$$

We seek steadystate axisymmetric solutions of this system. With this end in view, we introduce the Stokes potential ψ for a cylindrical set of coordinates r, φ, z:

$$r\, v_r = \partial_z \psi, \quad r\, v_z = -\partial_r \psi, \quad r\, v_\varphi = f(\psi), \tag{7.3}$$

where f is an arbitrary function.

Substituting (7.3) into (7.1), (7.2), we obtain

$$\tilde{\Delta}\psi = -ff' - r^2 F'(\psi), \quad \tilde{\Delta} \equiv r\partial_r r^{-1}\partial_r + \partial_z^2, \tag{7.4}$$

$$p/\rho = F(\psi) - \psi^2/2, \tag{7.5}$$

where $F(\psi)$ is an arbitrary function. This equation is known as the Grad–Shafranov equation (Shafranov, 1966). Because f and F are arbitrary, it has widely diverse solutions. By way of illustration, we quote one that is expressible in terms of elementary functions.

Let us define a sphere of radius a. The arbitrary functions are chosen inside it in such a way as to reduce (7.4) to the form

$$\tilde{\Delta}\psi + k_0^2\psi = r^2 c, \tag{7.6}$$

where k_0 and c are constants. The solution of (7.6) that is bounded at zero has the form

$$\psi = [A\varphi(R) + c/k_0^2]r^2, \tag{7.7}$$

where

$$\varphi = \frac{\sin k_0 R}{R^3} - \frac{k_0 \cos k_0 R}{R^2}, \quad R^2 = r^2 + z^2. \tag{7.8}$$

Here A is the soliton amplitude. Let the soliton be localized completely inside a sphere of radius a. For the soliton edge to be smooth one should set

$$\psi a = 0, \quad (\partial_R \psi)_a = 0.) \tag{7.9}$$

at $R \geq a$.These requirements establish the following relation between the constants in (7.7):

$$c = -k_0^2 A\varphi(a), \tag{7.10}$$

$$\tan k_0 a = 3k_0 a/(3 - k_0^2 a^2). \tag{7.11}$$

The first roots of this equation are $k_0 a = 5.76$ and 9.1. This solution describes a vortex that has all the three velocity components and is at rest relative to the medium. The toroidal component of the vorticity $\tilde{\Delta}\psi$ has a finite jump at the vortex boundary. Of the vortices travelling relative to the medium, Hill's vortex is the best known. It corresponds to the following solution of (7.4). We put $F = -(3u/a^2)\,\psi$ inside the sphere of radius a and $F = 0$ outside it. The toroidal component of velocity is everywhere zero. This corresponds to $f = 0$. Equation (7.4) then reduces to

$$\tilde{\Delta}\psi = \begin{cases} 3ur^2/a^2, & R \leq a, \\ 0, & R > a. \end{cases} \tag{7.12}$$

This has the solution

$$\psi = \begin{cases} -\dfrac{3}{4}\, ur^2 \left(1 - \dfrac{R^2}{a^2}\right), & R \leq a, \\ \dfrac{1}{2}\, ur^2 \left(1 - \dfrac{a^3}{R^3}\right), & R \geq a, \end{cases} \tag{7.13}$$

where u is the velocity of the vortex relative to the medium.

We have $v_z = u$ far from the vortex. Note that the perturbed velocity falls off as $1/R^2$ with distance from the vortex. All particles inside the sphere are trapped, that is, they move with the vortex. Hill's solution has two independent parameters: the velocity u and size a. A generalization to the case of a velocity $v_\varphi \neq 0$ is possible. A third independent parameter then appears in the solution. To determine it, we must match the solution of (7.6) to the solution of (7.12) for $R \geq 0$, thus fixing the coefficients A and c:

$$A = \frac{3}{2}\frac{u}{a\,\varphi'(a)}, \quad c = -\frac{3}{2}\frac{k_0^2\, u\, \varphi(a)}{\varphi'(a)}. \tag{7.14}$$

We have thus obtained an extension of Hill's solution that has a nonzero toroidal velocity. It has three free parameters k_0, u, a. Experiments with such vortices made by Stepanyants and his associates show that Hill vortices are stable, while travelling vortices having a toroidal velocity component are unstable. The stability of these vortices can be investigated by the Lyapunov method (see Appendix 2). To do this, we must have a sufficiently broad range of first integrals of motion for Euler's equations (7.1), (7.2). Note that, besides the well-known integrals of energy (E), momentum (**P**) and angular momentum (**M**),

$$E = \int (v^2/2) d^3x,$$

$$\mathbf{P} = \int \mathbf{v} d^3x, \tag{7.15}$$

$$\mathbf{M} = \int (\mathbf{r} \times \mathbf{v}) d^3x,$$

they also conserve the helicity

$$I = \int_V \mathbf{v} \cdot \text{curl } \mathbf{v} d^3x, \tag{7.16}$$

where V is an arbitrary volume within the surface on which $\mathbf{n} \cdot \text{curl } \mathbf{v} = 0$ and $\mathbf{n}$ is the normal vector at the surface. This can be generated in the following way: if a vector function $\mathbf{u}$ satisfies the equation

$$\partial_t \mathbf{u} + \nabla(\mathbf{u} \cdot \mathbf{v}) = \mathbf{v} \times \text{curl } \mathbf{u},$$

then the helicity density, defined as $h = \mathbf{u} \cdot curl\, \mathbf{v}$, is conserved locally, i.e. $\partial_t h + \text{div } h\mathbf{v}$, and if $\text{div } \mathbf{v} = 0$. Then a functional series of the first integrals

$$I_f = \int f(h)\, d^3x$$

exists, where f is an arbitrary function.

7.2. TRANSIENT 2D SOLUTIONS IN LAGRANGE VARIABLES

The above steady-state solutions of the equations of ideal hydrodynamics were obtained under the assumption of axial symmetry. Besides these solutions, non-steady ones also exist. They are naturally much harder to obtain, so that few solutions of this sort are available at present. Abrashkin and Yakubovich (1984) have found fairly general 2D solutions of (7.1) and (7.2). These are conveniently derived using Lagrange variables. Let x_0, y_0 be the initial coordinates of a fluid particle. Then its coordinates x, y at time t are functions of x_0, y_0, t. The pressure gradient accelerates them:

$$\ddot{x} = -\rho^{-1}\partial_x p, \quad p = p(x, y, t), \quad \rho = \text{const}, \tag{7.17}$$

$$\ddot{y} = -\rho^{-1}\partial_y p, \quad x = x(x_0, y_0, t), \quad y = y(x_0, y_0, t). \tag{7.18}$$

To eliminate the derivatives with respect to x and y from the right-hand side, we multiply these equations by $\partial_{x_0} x$ and $\partial_{x_0} y$, respectively, and add them. Then multiplying them by $\partial_{y_0} x$ and $\partial_{y_0} y$ and adding, we obtain

$$\ddot{x}\partial_{x_0} x + \ddot{y}\partial_{x_0} y = -\rho^{-1}\partial_{x_0} p, \tag{7.19}$$

$$\ddot{x}\partial_{y_0} x + \ddot{y}\partial_{y_0} y = -\rho^{-1}\partial_{y_0} p. \tag{7.20}$$

When written in this form, p can be regarded as a function of x_0, y_0, t, in contrast to (7.17) and (7.18). We integrate this system from zero to t. Then, for instance, for (7.19), we obtain

$$\int_0^t (\ddot{x}\partial_{x_0} x + \ddot{y}\partial_{x_0} y)\, dt = \dot{x}\partial_{x_0} x + \dot{y}\partial_{x_0} y - v_{0x}$$
$$- \frac{1}{2}\partial_{x_0}\int_0^t [(\dot{x})^2 + (\dot{y})^2]dt = -\rho^{-1}\partial_{x_0}\int_0^t p\,dt, \tag{7.21}$$

where v_{0x} is the initial value of the velocity component along x at time $t = 0$. In a similar fashion, (7.20) yields

$$\int_0^t (\ddot{x}\partial_{y_0} x + \ddot{y}\partial_{x_0} y)dt = \dot{x}\partial_{y_0} x + \dot{y}\partial_{y_0} y - v_{0y}$$
$$- \frac{1}{2}\partial_{y_0}\int_0^t [(\dot{x})^2 + (\dot{y})^2]dt = -\rho^{-1}\partial_{y_0}\int_0^t p\,dt. \tag{7.22}$$

We now differentiate (7.21) with respect to y_0, (7.22) with respect to x_0 and subtract the results. We get

$$\{x, \dot{x}\} + \{y, \dot{y}\} = -\Omega, \tag{7.23}$$

where $\{A, B\} \equiv \partial_{x_0} A \partial_{y_0} B - \partial_{x_0} B \partial_{y_0} A$ and Ω is the initial value of the vorticity ($\Omega \equiv \mathrm{curl}_z v_0$).

It is readily shown that the incompressibility condition (7.2) can be written in the Lagrange variables x_0, y_0 as the condition that an elementary volume of incompressible fluid be conserved,

$$\{x, y\} = 1 . \tag{7.24}$$

Equations (7.1) and (7.2) thus reduce to (7.23) and (7.24). These can be simplified by transforming to the complex variables $w = x + iy$, $\eta = x_0 + iy_0$. Here w is not an analytic function, that is, it depends both on η and on its complex conjugate η^*. The equations of ideal hydrodynamics in these variables are written as follows:

$$\{\dot{w}, w^*\} = -\Omega, \quad \Omega = \Omega(\eta, \eta^*), \tag{7.25}$$

$$\{w, w^*\} = 1, \quad w = w(\eta, \eta^*, t), \tag{7.26}$$

where $\{A, B\} \equiv \partial_\eta A \partial_{\eta^*} B - \partial_\eta B \partial_{\eta^*} A$ and w is the complex velocity.

Abrashkin and Yakubovich (1984) have found the following solution for these equations:

$$w = G(\eta) \exp(i\lambda t) + F(\eta^*) \exp(i\mu t), \tag{7.27}$$

where λ and μ are arbitrary real numbers and G and F are analytic functions related by a condition that follows from (7.26):

$$|\partial_\eta G|^2 + |\partial_{\eta^*} F|^2 = 1. \tag{7.28}$$

This solution oscillates at the two frequencies λ and μ . It can be shown that the corresponding motion for fluid particles is the superposition of two circular motions at the frequencies λ and μ . According to (7.25), the vorticity Ω in this solution is

$$\Omega = 2(\lambda |\partial_\eta G|^2 - \mu |\partial_{\eta^*} F|^2). \tag{7.29}$$

When expressed in Euler variables, both the velocity and the vorticity corresponding to these equations depend on the time and coordinates in a complicated manner. We now consider whether the above solution can be ascribed to a potential velocity field. The latter is conveniently represented in Euler's variables. The dependence of the complex velocity $\dot{w} = v_x + iv_y$ on the complex coordinate $w = x + iy$ and the time will be specified in parametric form using a complex parameter ξ :

$$\dot{w} = i\lambda G_1(\xi^*) \exp(i\lambda t) + i\mu F_1(\xi^*) \exp(i\mu t), \tag{7.30}$$

$$w = G_1(\xi) \exp(i\lambda t) + F_1(\xi) \exp(i\mu t). \tag{7.31}$$

It is readily verified that the vorticity and divergence in this solution are zero, because $\dot{w}$ is a function of ξ and t only. This can be seen from the general relations div v $= 2\mathrm{Re}\partial_\xi \dot{w}$ and $\Omega = -2\mathrm{Im}\partial_\xi \dot{w}$. Let us assume that the vorticity at the initial time occupies a region whose boundary is given by $\eta = \eta_{bound}$ (which corresponds to the boundary equation $y_0 = y_0(x_0)$). It is easy to see that then the solution (7.30) and (7.31) at an arbitrary instant of time transforms continuously into (7.27) at the values $\xi = \xi_{bound}$ determined from

$$\begin{aligned} G(\eta_b) &= G_1(\xi_b), \quad F(\eta_b^*) = F_1(\xi_b), \\ G(\eta_b) &= G_1(\xi_b^*), \quad F(\eta_b^*) = F_1(\xi_b^*). \end{aligned} \tag{7.32}$$

From this one can see that ξ_{bound} is purely real. The range of variation of ξ is chosen so that the coordinates w of the solenoidal and the potential regions should not overlap. The relations (7.32) define $F_1(\xi)$ and $G_1(\xi)$ on the complex ξ-plane as analytic continuations of G and F found from their values at the contour $\xi = \xi_{bound}$.

As an example, let us take the expression for the solenoidal region in the form

$$w = c\zeta \exp(i\lambda t) + F(\zeta^*), \quad \zeta \equiv \exp(ik\eta). \tag{7.33}$$

Suppose the vortex lies within the unit circle $|\zeta| = 1$. The expression (7.33) corresponds to the choice $G(\eta) = \exp(ik\eta)$ and $\mu = 0$. The potential flow in the outer region of this circle can be written as

$$w = c\,\zeta \exp(i\lambda t) + F(\zeta^{-1}),$$

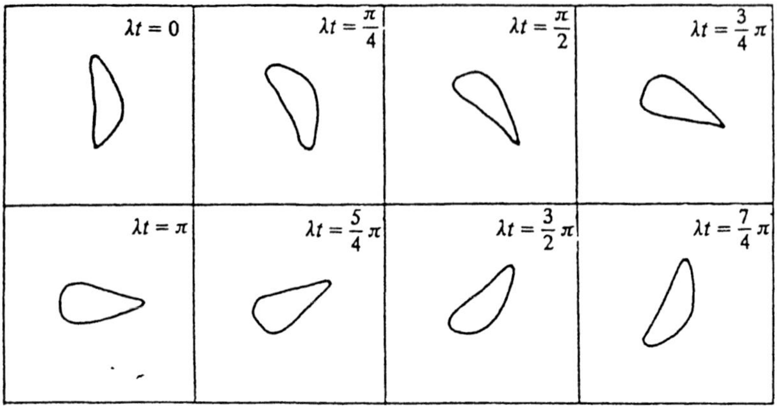

Figure 7.1. Evolution of the boundary of a vortex spot with a potential flow around it.

$$\dot{w} = i\lambda c(\zeta^*)^{-1} \exp(i\lambda t). \tag{7.34}$$

This flow coincides with the flow around a point vortex at infinity. Figure 7.1 shows the behaviour of the boundary of the solenoidal region for the case $F(\zeta^*) = \alpha\zeta^* + \beta(\zeta^*)^2$, where α and β are real-valued parameters.

7.3. SOLITARY VORTICES IN MAGNETOHYDRODYNAMIC THEORY

Solutions in the form of toroidal equilibrium configurations are of interest in theoretical studies relating to plasma confinement in magnetic confinement devices. The plasma pressure in these solutions is greatest at the centre. For this reason, as shown by Shafranov (1966), they cannot be solitary. Petviashvili, Pokhotelov and Chudin (1982) and Petviashvili, Pokhotelov and Stenflo (1986) have found solutions to the magnetohydrodynamic equations in the form of solitary toroidal vortices. The study of toroidal vortices has become especially important in connection with the development of powerful quasistationary plasma-static and plasma-dynamic confinement systems in which hydrodynamic motion coexists with a solenoidal magnetic field (Morozov and Solov'ev, 1980).

Solitary MHD vortices can occur in space plasmas, for example, in the solar wind which, since it is large and comparatively homogeneous, is a convenient object for studying Alfvén waves. Measurements show that the wind velocity near the Earth is $n \simeq 5\times10^5$ m/s, the Alfvén velocity is $c_A \simeq 5\times10^4$ m/s, the plasma density perturbation in the observed turbulence is negligibly small, and the fluctuations in the magnetic field components are on the order of the average field, while those in the plasma velocity are on the order of order c_A (Belcher and Davis, 1971; Coleman, 1968; Dobrovolny *et al.*, 1980). This indicates that strong Alfvén turbulence is present in the solar wind. Observations reveal an inertial range for the wavenumbers $k > k_0 \simeq 10^{-8}\,\mathrm{m}^{-1}$ where the turbulence power spectrum has the form $W_k \propto k^{-1.6}$. The application of Kolmogorov dimensional analysis to Alfvén turbulence is based on the assumption of an isotropic power spectrum. Kraichnan (1965) used this assumption to demonstrate that the Alfvén turbulence spectrum obeys $W_k \propto k^{-1.5}$. Since the frequency spectrum of the Alfvén waves is strongly anisotropic, the assumption of isotropic turbulence requires substantiation. In addition, the question remains as to why W_k is weakly dependent on k for $k < k_0$. In the following we show that Alfvén turbulence may consist of structures with the form of toroidal solitons travelling along the magnetic field at the Alfvén velocity. Some of the magnetic surfaces in these solitons may be closed and have the form of toroids, making the soliton topologically stable. In that case, $1/k_0$ represents the size of the largest solitons in the turbulence. A soliton of size $1/k_0$ in the wavenumber range $k < k_0$ has a nearly constant power spectrum which falls off rapidly for $k > k_0$. The rapid fall-off in the spectrum of the observed turbulence for $k > k_0$ can be accounted for by the presence of solitons of size smaller than $1/k_0$ with random distributions of size and amplitude. Isotropic turbulence emerges as a consequence of the soliton spectrum being approximately isotropic. Similarly sized solitons interact strongly. Colliding solitons produce a cascade of energy Γ_k towards greater k. Since the solitons are purely topological and are independent of any specific size or frequency, the cascade for $k > k_0$ cannot be much different from the turbulence cascade found by Kraichnan (1965). The small difference between the observed and theoretical spectrum noted by Belcher and Davis (1971) and Coleman (1968) for $k > k_0$ can be explained by the extra attenuation introduced by Kadomtsev and Petviashvili (1973).

As we shall see later on, MHD vortices are completely localized, if they travel at the Alfvén velocity in the ambient magnetic field. If, on the other hand, they travel at a velocity different from the Alfvén velocity, then they may be localized according to a power law. Exactly as in ordinary hydrodynamics, MHD vortices can only have a poloidal component or all three components of velocity and the magnetic field. Perturbations in the magnetic field and in the velocity are parallel to one another, as in the Alfvén wave case, except for some cases of which special mention will be made. For this reason, these vortices can be referred to as Alfvén vortices. Plasma compressibility is insignificant here.

We proceed from the equations of ideal magnetic hydrodynamics assuming an incompressible plasma:

$$\text{div } \mathbf{v} = \text{div } \mathbf{B} = 0, \tag{7.35}$$

$$\partial_t \mathbf{v} + \mathbf{v}\times\text{curl } \mathbf{v} - (\mathbf{B}\times\text{curl } \mathbf{B})/4\pi\rho = -\nabla(p/\rho + v^2/2), \tag{7.36}$$

$$\partial_t \mathbf{B} = \text{curl}(\mathbf{v}\times\mathbf{B}). \tag{7.37}$$

We shall find the simplest analytical expression for a toroidal soliton. We seek a solution that travels at the Alfvén velocity. Let the magnetic field consist of a constant part $\mathbf{B}_0$ and a soliton part $\mathbf{B}_1$. Then in a reference frame travelling along $\mathbf{B}_0$ at velocity $c_A = B_0/(4\pi\rho)^{1/2}$, (7.36) yields

$$p + \rho v^2/2 = \text{const}. \tag{7.38}$$

where $\mathbf{v} = \mathbf{c}_A + \mathbf{v}_1$, $\mathbf{v}_1 = \mathbf{B}_1/(4\pi\rho)^{1/2}$ and $\mathbf{B}_1 = \mathbf{B}_1(x, y, z - c_A t, t)$.

Let us introduce the Stokes potential

$$r\mathbf{B}_1 = \mathbf{e}\times\nabla\psi + \mathbf{e}k_0\psi, \tag{7.39}$$

where $\mathbf{e}$ is the unit vector in the azimuthal direction in a cylindrical coordinates system and k_0 is an arbitrary constant proportional to the toroidal component of the magnetic field. We require the soliton to consist of toroidal surfaces ψ = constant. This leads to the condition

$$\mathbf{B}_1\times\text{curl } \mathbf{B}_1 = c\nabla\psi, \tag{7.40}$$

where c is an arbitrary constant.

Substituting (7.39) into (7.40) and integrating, we obtain equation (7.6) which has been derived in ordinary hydrodynamics. Hence, we conclude that Alfvén vortices travelling at velocity c_A are identical to solitary vortices at rest with respect to the medium in ordinary hydrodynamics.

We now discuss a broader class of toroidal Alfvén solitons whose shape can be found by numerical methods. To do this, we express the soliton magnetic field in terms of the Stokes potential as

$$r\mathbf{B}_1 = \mathbf{e}\times\nabla\psi + \mathbf{e}f(\psi), \tag{7.41}$$

where $f(\psi)$ is an arbitrary function of ψ.

The condition for the existence of magnetic surfaces is expressed in the form

$$\mathbf{B}_1\times\text{curl}\,\mathbf{B}_1 = \nabla F(\psi), \tag{7.42}$$

where $F(\psi)$ is an arbitrary function. Substituting (7.41) into (7.42), we obtain the Grad-Shafranov equation (7.4).

Petviashvili, Pokhotelov and Chudin (1982) have shown that (7.4) has solitary smooth solutions, provided it can be reduced to the form

$$\tilde{\Delta}\psi = A^2r^2\psi - B\psi^n, \tag{7.43}$$

where n is an integer that characterizes the degree of nonlinearity and A and B are constants that determine the soliton amplitude and size.

When $n = 2$, the vorticity in a torus is the most compact (the vorticity approaches the z-axis). As n increase, the vorticity moves away from the z-axis and solutions that are odd with respect to z are possible for odd n. The most compact of the antisymmetric solutions is the one for $n = 3$.

With the aid of a similarity transformation, equation (7.43) can be written in the dimensionless form

$$\Delta^*\varphi = \rho^2\varphi - \varphi^n\,, \quad \Delta^* \equiv \rho\partial_\rho\rho^{-1}\partial_\rho + \partial_\zeta^2\,, \tag{7.44}$$

where $\rho = A^{1/2}r$, $\zeta = A^{1/2}z$, $\varphi = (B/A)^{1/(n-1)}\psi$.

Contours of the solution of (7.44) for $n = 2$ and $n = 3$ are shown in Figures 7.2, 7.3. Figures 7.4, 7.5 show contours of the toroidal component of the current which is proportional to $\tilde{\Delta}\psi$. One can see that the current moves away from the axis of symmetry when $n = 3$.

Exactly the same solutions were obtained for vortices at rest where the plasma velocity is everywhere zero, while the magnetic field is given by (7.39) or (7.41).

When a stochastic set of these solitons exists, turbulence can develop in which the energy cascades toward greater wavenumbers. Note that if

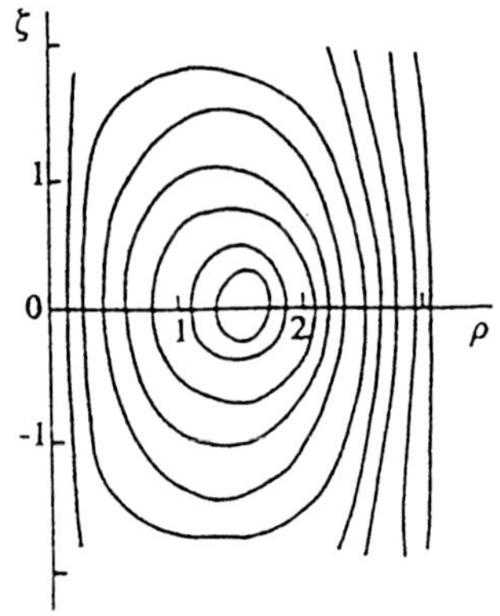

Figure 7.2. Contours of the Stokes potential φ of the soliton solution of (7.44) with a quadratic nonlinearity (the maximum value is reached at $\rho = 1.6$ and is equal to 3.8).

the turbulence consists of the Alfvén toroidal solitons considered above, then it must be almost isotropic. Also, Petviashvili, Pokhotelov, and Stenflo (1986) have pointed out that when discontinuous derivatives appear at the wave crests, the harmonics are attenuated through absorption of their energy at the discontinuities. This occurs, for example, in acoustic turbulence and in the turbulence of gravity waves on the ocean

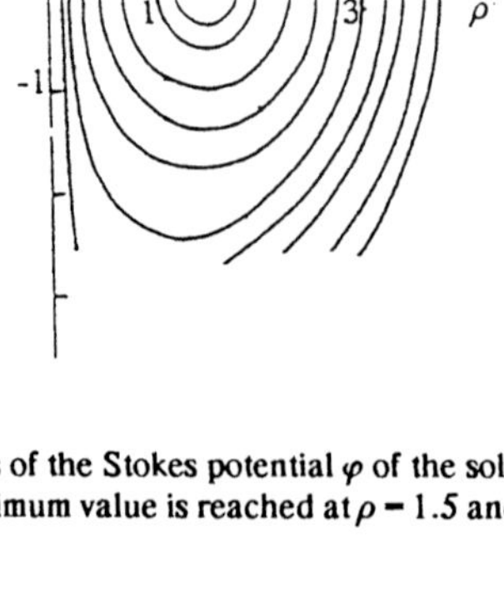

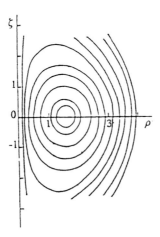

Figure 7.3. Contours of the Stokes potential φ of the soliton solution of (7.44) with a cubic nonlinearity (the maximum value is reached at $\rho = 1.5$ and is equal to 3.3).

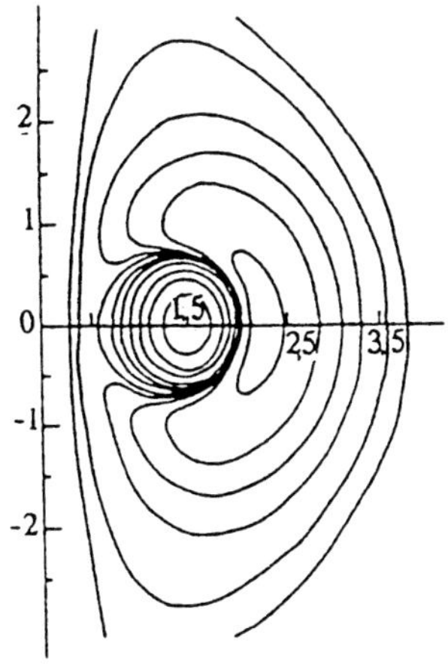

Figure 7.4. Contours of the toroidal component of the vorticity (of the electric current) in equation (7.44) for $n = 2$ (the minimum value is reached at $\rho = 1.5$ and is equal to -29, the maximum, at $\rho = 2.3$, is equal to 1.9).

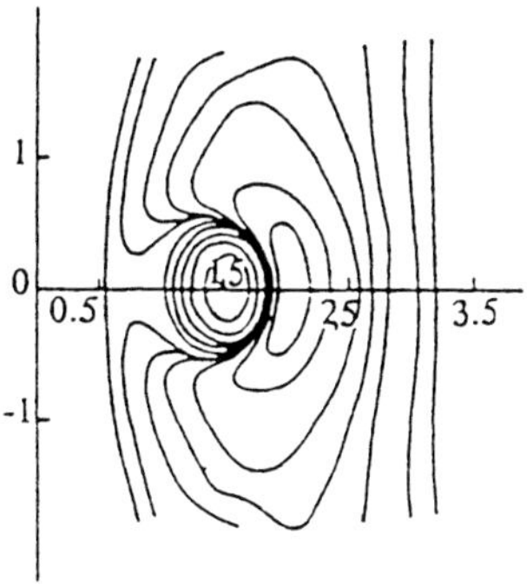

Figure 7.5. Contours of the toroidal component of the vorticity (of the electric current) in equation (7.44) for $n = 3$ (the minimum value is reached at $\rho = 1.5$ and is equal to -43, the maximum at $\rho = 2.0$ and is equal to 6.1).

surface. The appearance of discontinuities means that there is a phase correlation between harmonics with different wavenumbers. A similar correlation can also occur in media where stable structures can exist in the form of solitons or solitary vortices. The effect can be taken into account by inserting an extra term in the equation for the energy flow along the spectrum in a way similar to that used by Petviashvili, Pokhotelov and Stenflo (1986)

$$\partial_t W_k + \partial_k \Gamma_k = -\alpha \Gamma_k / k, \tag{7.45}$$

where W_k is the spectral energy density, Γ_k the energy flux (cascade) along the spectrum, and α a constant.

The right-hand side in (7.45) was derived by Kadomtsev and Petviashvili (1973) from dimensional considerations for the case of acoustic turbulence. Here, it describes the phase correlation between harmonics with different wavenumbers in solitons with discontinuous vorticity.

The expression for Γ_k is found from (7.35)–(7.37) by dimensional considerations $\Gamma_k \propto k\,\omega_k W_k$, where ω_k is a characteristic frequency which can be estimated using (7.35)–(7.36). These yield $\omega_k \propto k^2 W_k / c_A$. $\Gamma_k =$ const for the Kolmogorov turbulence. For soliton turbulence in the steady-state case, we have $\partial_t W_k = 0$. Hence, (7.45) gives $k^\alpha \Gamma_k =$ constant, so that $W_k \propto k^{-(3+\alpha)/2}$. Determining α requires knowledge of the soliton interaction, which is a difficult problem. For this reason, we shall use experimental evidence, as Kadomtsev and Petviashvili (1973) did. According to measurements in the solar wind (Belcher and Davis, 1971), Coleman (1968) and Dobrovolny *et al,* 1980), we have $W_k \propto k^{-\sigma}$ ($\sigma = 1.5 - 1.6$) at the level of the Earth's orbit. Hence we conclude that $\alpha = 0 - 0.2$ for Alfvén turbulence.

7.4. MODEL EQUATION FOR ELECTRON HYDRODYNAMICS

The ion motion in plasma oscillations can be neglected in a number of cases. In particular, this can be done for propagation of electromagnetic waves at a frequency above the ion cyclotron frequency and a wavelength shorter than c/ω_{Pi}. In that case the ions provide a background to ensure plasma quasineutrality.

The equations of motion for the electrons are

$$m_e d_t \mathbf{v} = -e\mathbf{E} - \nabla p - (e/c)(\mathbf{v}\times\mathbf{B}), \tag{7.46}$$

$$\operatorname{div} \mathbf{v} = 0.$$

We shall express the electric and magnetic field in terms of the scalar and vector potentials φ and $\mathbf{A}$ as

$$\mathbf{E} = -\nabla\varphi - (1/c)\partial_t \mathbf{A}, \quad \mathbf{B} = \text{curl } \mathbf{A}.$$

Then, taking the curl of (7.46) and introducing the generalized momentum $\mathbf{p} = m_e\mathbf{v} - (e/c)\mathbf{A}$, we obtain

$$\partial_t \text{curl } \mathbf{p} = \text{curl}(\mathbf{v}\times\text{curl } \mathbf{p}). \tag{7.47}$$

From (7.47) one can see that the generalized momentum is frozen into the electron fluid. In the following we restrict ourselves to considering rather dense plasmas $\omega_{Pe} >> \omega_{Be}$ and the frequency range $\omega_{Bi} \ll \omega \ll \omega_{Pe}$. The displacement current can then be neglected, and the Maxwell equation has the form

$$\text{curl } \mathbf{B} = \frac{4\pi}{c}\mathbf{j} \simeq -\frac{4\pi en}{c}\mathbf{v}. \tag{7.48}$$

Here it has been assumed that the ion contribution to the electric current is negligibly small. The density of the plasma particles is taken to be constant. Changes in the density are mainly due to RF pressure effects. However, this effect is known to occur in the approximation which is cubic in the pertubation amplitude. Here we are considering nonlinear effects that are quadratic in the amplitude. For this reason, density perturbations will be disregarded in the following. We use (7.48) to eliminate the velocity $\mathbf{v}$ from (7.47). This yields the main equation of electron hydrodynamics:

$$\partial_t\Omega = (e/m_e c)r_0^2 \text{ curl } (\Omega\times\text{curl } \mathbf{B}), \tag{7.49}$$

$$\Omega \equiv \mathbf{B} - r_0^2\nabla^2\mathbf{B},$$

where $r_0 = c/\omega_{Pe}$ is the skin depth. If we introduce the typical magnitude of the magnetic fields $\mathbf{B}_0$, then (7.49) can be written in the dimensionless form

$$\partial_t\mathbf{q} = \text{curl}(\mathbf{q}\times\text{curl } \mathbf{H}), \quad \mathbf{q} \equiv \mathbf{H} - \nabla^2\mathbf{H}. \tag{7.50}$$

Here $\mathbf{H} = \mathbf{B}/B_0$, time is measured in units of $m_e c/eB_0$, and the coordinates in units of r_0. This equation has integrals of energy, momentum and angular momentum

$$W = \frac{1}{2}\int \mathbf{H}\cdot\mathbf{q}d^3x,$$

$$\mathbf{P} = \frac{1}{2}\int (\mathbf{r}\times\mathbf{q})d^3x, \tag{7.51}$$

$$\mathbf{M} = \frac{1}{3}\int (\mathbf{r}\times(\mathbf{r}\times\mathbf{q}))d^3x.$$

It also conserves the flux of $\mathbf{q}$ across an arbitrary closed fluid contour.

It is of interest to extend the concept of helicity to electron hydrodynamics. This can be done by defining a vector $\mathbf{Q}$, an analogue of the magnetic vector potential $\mathbf{A}$, and the electron helicity s, an analogue of MHD helicity. Let $\mathbf{q} = \text{curl}\,\mathbf{Q}$, $s = \mathbf{Q}\cdot\mathbf{q}$. Then from (7.50), we have

$$\partial_t\mathbf{Q} = \mathbf{v}\times\mathbf{q} + \nabla\psi, \tag{7.52}$$

where ψ is an arbitrary function. The choice of ψ determines the gauge for $\mathbf{Q}$. We introduce the following gauge

$$\psi = -\mathbf{Q}\cdot\mathbf{v}. \tag{7.53}$$

Then (7.50), (7.52), and (7.53) give

$$\partial_t s + \text{div}\, s\,\mathbf{v} = 0. \tag{7.54}$$

This equation shows that the electron helicity is conserved locally at each point. The consequences of this equation include some new conserved integrals. Recalling that div $\mathbf{v} = 0$ in electron hydrodynamics, we find that

$$I = \int_D f(s)\, d^3x, \tag{7.55}$$

is conserved. Here f is an arbitrary function, D is an arbitrary fluid volume whose points are moving at velocity $\mathbf{v}$. This integral can be used to prove the stability of electron structures by the Lyapunov method. In the linear approximation, (7.50) describes the propagation of helicons with a dispersion relation given by

$$\omega = \pm\frac{k(\mathbf{k}\cdot\mathbf{H}_0)}{1+k^2}. \tag{7.56}$$

These waves are frequently observed in the Earth's magnetosphere, where they are generated as a result of the cyclotron instability. They also arise as a result of electrical discharges in the atmosphere. They are then called whistlers or helicons. Helicons are also observed in semiconductor plasmas. In the limit of large wavenumbers, helicons transform into electron cyclotron waves with a frequency approaching ω_{Be}.

7.5. TOROIDAL ELECTRON VORTICES

We now show that a helicon wave packet can travel along the axis of a constant helical magnetic field in the form of a toroidal vortex that is localized exponentially (Marnachev, 1987; Isichenko and Marnachev, 1987). The solution is only slightly dependent on the helical pitch, which can range between zero (Z-pinch) and infinity (straight field). We seek a solution to (7.50) in the form $\mathrm{H} = \mathrm{H}(r, z - ut)$, where r is the radial coordinate in a cylindrical coordinates sytem. From (7.50) we then obtain

$$\mathrm{q}\times(\mathrm{curl}\ \mathrm{H} + u\mathrm{e}_z) = \nabla\Phi, \tag{7.57}$$

where Φ is an arbitrary function and e_z is the unit vector in the z direction. We express H in terms of ψ and f, similarly to (7.41). From (7.57) we then have

$$\tilde{\Delta}\psi - \psi = F, \quad F = F(\tilde{f}), \tag{7.58}$$

and

$$\tilde{\Delta} f - f + F'\psi = -FF' + r^2\Phi f', \tag{7.59}$$

where F and Φ are arbitrary functions of $\tilde{f} \equiv f + ur^2/2$. We introduce a sphere of radius $R = R_0$, where $R^2 = r^2 + (z - ut)^2$. We set $F = b\tilde{f}$ and $\Phi' = a$ outside the sphere. Then for $R > R_0$ we have

$$\tilde{\Delta}\psi - \psi = b(f + ur^2/2), \tag{7.60}$$

and

$$\tilde{\Delta} f - f + b\psi = -b^2 f + r^2(a - ub^2/2). \tag{7.61}$$

This set of equations has solutions of the form

$$\psi = \left[\frac{A\chi(\varkappa_1 R)}{\varkappa_1^2 - 1} + \frac{A^*\chi(\varkappa_2 R)}{\varkappa_2^2 - 1}\right] br^2 - b(u/2 - a)r^2 \tag{7.62}$$

and

$$f = [A\chi(\varkappa_1 R) + A^*\chi(\varkappa_2 R)]r^2 - ar^2. \tag{7.63}$$

Here $\chi(x) = x^{-3/2}K_{3/2}(x)$, $K_{3/2}$ is the McDonald function, and $\varkappa_{1,2}$ are the roots of the dispersion relation

$$(\varkappa^2 + b^2/2 - 1)^2 = b^2(b^2/4 - 1), \tag{7.64}$$

obtained by substituting (7.62) and (7.63) into (7.60) and (7.61). They have the form

$$\varkappa_{1,2}^2 = 1 - b^2/2 \pm (b^2/4 - b^2)^{1/2}. \tag{7.65}$$

For χ to vanish at infinity the roots (7.65) must have real parts. This only occurs when $b^2 < 4$. Then as $R \to \infty$, we have: $\psi \to b(a - u/2)r^2$ and $f \to -ar^2$. This means that the magnetic field far from the vortex has the components

$$H_{0z} = (u - 2a)b \quad \text{and} \quad H_{0\varphi} = -a\,r. \tag{7.66}$$

There is also an effective radial electric field equal to $E_r = -\partial_z\Phi = a(u - 2a)$ (in units of B_0). Inside the sphere F and Φ are chosen so that (7.58) and (7.59) become

$$\tilde{\Delta}\psi - \psi = c\,(f + u\,r^2/2) \tag{7.67}$$

and

$$\tilde{\Delta}f - f + c^2\psi = -c^2 f + r^2(d - c^2u/2), \tag{7.68}$$

where c and d are constants. This system has the following solutions

$$\psi = -c\left[\frac{D_1}{k_1^2 + 1}\frac{j_1(k_1 R)}{R} + \frac{D_2}{k_2^2 + 1}\frac{j_1(k_2 R)}{R} + \frac{u}{2} - d\right] r^2 \tag{7.69}$$

and

$$f = [D_1 j_1(k_1 R) + D_2 j_1(k_2 R) - Rd](r^2/R), \tag{7.70}$$

where

$$k_{1,2}^2 = -1 + c^2/2 \pm (c^2/4 - c^2)^{1/2}, \tag{7.71}$$

are the roots of the dispersion relation and $j_1(\zeta)$ is the spherical Bessel function of the first kind. These solutions inside and outside the sphere must be matched at the boundary $R = R_0$ subject to the condition $\tilde{f} = f + ur^2/2 = 0$ and the continuity of $\psi, f, \nabla\psi$ and ∇f. The magnetic field and the electron speed are continuous at the boundary, while the z-component of the velocity vorticity $\tilde{\Delta} f/r$ is discontinuous.

Isichenko and Marnachev (1987) derived these solutions for a constant magnetic field with $H_{0\varphi} = 0$. Marnachev (1987) has investigated them for the case $H_{0z} = 0$ in a dc field with $j_{0z} = \text{const}$. This solution is of interest in that it can occur in the central part of a Z-pinch, where there is no longitudinal magnetic field. They have probably been observed as luminous spots at the terminal stage of high-power electric discharges in Z-pinches (Filippov, 1983). Numerous heavy impurities appear in a discharge at this stage and produce X-ray emission. It has been repeatedly pointed out in the literature that the radiation is emitted from isolated regions with very small dimensions on the order of r_0, rather than from the whole volume. If one tried to explain this radiation under the assumption that the plasma is in thermodynamic equilibrium, then one has to assume that the particle density in the luminous spots is several orders of magnitude greater than the density of a solid. Another explanation is as follows: as a result of instabilities, toroidal vortices of the type described above of size $\sim r_0$ form in the plasma. The electrons trapped in these vortices are far from thermodynamic equilibrium. It is possible that the growth or decay of such vortices accompanied by the appearance of an electric field along B produces runaway electrons with energies far above those of the background electrons. A small number of these electrons could collide with heavy impurities and produce X-ray emission of the same intensity as from a high-density plasma with a Maxwellian velocity distribution.

Note that when $B_{0z} = 0$, the propagation velocity of the vortex coincides with the unperturbed current velocity of the electrons; that is, in this case the vortex is at rest relative to the background electrons. For the equation of electron hydrodynamics (7.50) to be valid under these conditions, the electron thermal speed must be small compared to the current velocity in the vortex.

7.6. TOPOLOGICAL SOLITONS IN MEDIA WITH CONSTANT MAGNETIZATION

Unique waves can exist in substances with a constant magnetization vector **M** (that is, in ferromagnetic and antiferromagnetic substances). In these substances, **M** oscillates without varying in magnitude. The curl and divergence of this vector are generally nonzero. The equation for these waves was derived by Landau and Lifshits (Kosevich, Ivanov and Kovalev, 1983). In the simplest case it has the form

$$\partial_t \mathbf{m} = -\mathbf{m}\times\Omega - \gamma\mathbf{m}\times(\mathbf{m}\times\Omega), \tag{7.72}$$

$$\Omega = \omega_{\mathrm{H}} + \delta E/\delta\mathbf{m}. \tag{7.73}$$

Here **m** is the unit vector parallel to **M**, ω_{H} is the spin magnetic frequency vector in a constant magnetic field with intensity H, E is the energy of the substance, and γ is the relaxation constant. The unit vector **m** can be expressed in terms of two functions θ and φ of the time and position in accordance with the formulas

$$m_x = \sin\theta\cos\varphi, \quad m_y = \sin\theta\sin\varphi, \quad m_z = \cos\theta. \tag{7.74}$$

Let us consider uniaxial crystals and make the z-axis point along the direction of the anisotropy. In this comparatively simple case, we have:

$$E = \frac{\omega_0}{2}\int w d^3x, \tag{7.75}$$

where

$$w = l^2[(\nabla\theta)^2 + (\nabla\varphi)^2\sin^2\theta] + b\sin^2\theta. \tag{7.76}$$

Here ω_0, l, b are constants. ω_0 is the frequency of the homogeneous ferromagnetic resonance, l is the magnetic length, b the anisotropy constant. If b is positive, then a state in which **m** is everywhere along z corresponds to the minimum energy. In this case one speaks of an anisotropy with an axis of slight magnetization. When $b<0$, we have an anisotropy with a plane of slight magnetization.

If (7.72) does not include dissipation ($\gamma = 0$), then, as can easily be seen, E is conserved. The other conserved quantities are the momentum, the projection of the magnetic moment onto the z-axis, and the angular momentum

$$\mathbf{P} = \int (1 - \cos\theta)\nabla\varphi d^3x, \tag{7.77}$$

$$N = \int (1 - \cos\theta) d^3x, \tag{7.78}$$

$$\mathbf{K} = \int (1 - \cos\theta)(\mathbf{r}\times\nabla\varphi) d^3x.$$

When we transform to the variables (7.74) and use (7.73) and (7.75), equation (7.72) takes the form (for simplicity we assume that $\gamma = 0$ and ω_{H} points along z)

$$\omega_0^{-1}(\partial_t\varphi - \omega_{\text{H}})\sin\theta + l^2\nabla^2\theta = [b + l^2(\nabla\varphi)^2]\sin\theta\cos\theta, \tag{7.80}$$

$$\omega_0^{-1}\partial_t\theta\sin\theta = l^2 \operatorname{div}(\sin^2\theta\nabla\varphi). \tag{7.81}$$

In the linear approximation equations (7.80) and (7.81) describe spin waves with the dispersion relation

$$\omega = \omega_0 + \omega_{\text{H}} + \omega_0 l^2 k^2, \qquad \theta = \text{const}, \quad \varphi = \omega t - \mathbf{kr}. \tag{7.82}$$

We now consider the simplest soliton solutions of (7.80) and (7.81). Special interest attaches to solutions that are independent of time, since they are not attenuated in the presence of dissipation, as can be deduced from (7.72). When $\partial_t = 0$, it follows from (7.72) and (7.73) that $\Omega = 0$, which is a necessary condition for minimum energy and stability. When expressed in terms of the variables θ, φ this condition can be derived from (7.80) and (7.81) for $\partial_t = 0$.

An important role is played in the physics of ferromagnets by solutions that are functions of a single coordinate, one perpendicular to the axis of anisotropy ($\theta = 0$). Equation (7.80) then becomes the Sine–Gordon equation. If we put φ = const and $\theta = \theta(x)$, then we obtain an equation for a steady-state 1D wave from (7.80) and (7.81):

$$l^2\partial_x^2\theta = (\omega_{\text{H}}/\omega_0 + b\cos\theta)\sin\theta. \tag{7.83}$$

If there is no external magnetic field ($\omega_H = 0$), then a solitary solution of (7.83) can be obtained analytically and has the form

$$\cos\theta = \tanh(x\sqrt{b}/l). \tag{7.84}$$

In this solution, the magnetization vector experiences a half rotation over a characteristic distance of $l/\sqrt{b}$. When boundaries are present, the most energetically favourable solution is one in the form of a periodic structure, a sign-varying sequence of domains, as shown by Landau and Lifshits. The concept of domains was introduced by Weiss in 1907.

We now consider 2D solutions that are independent of z and time, the so-called magnetic vortices (Kosevich, Ivanov and Kovalev, 1983). There is a change in the sign of m_z at some distance from the centre in these vortices, provided they are stable. The radial and azimuthal components of the magnetization form a logarithmic spiral whose parameters are difficult to find in analytic form. We put $\theta = \theta(r)$, $\varphi = \chi + \varphi_0$ in a cylindrical coordinate system r, χ, z, where φ_0 is a constant. Then (7.80) becomes

$$l^2 r^{-1}\partial_r r\,\partial_r\theta = \left[(b + l^2/r^2)\cos\theta + \omega_H/\omega_0\right]\sin\theta. \tag{7.85}$$

Kosevich, Ivanov and Kovalev (1983) have found a solitary solution to this equation. It has a negative magnetization along the external magnetic field at the vortex centre which becomes positive at the edges. The radial component of the magnetization is

$$m_r = m_x\cos\chi + m_y\sin\chi = \sin\theta\cos\varphi_0. \tag{7.86}$$

The azimuthal component is equal to $m_x = \sin\theta\sin\varphi_0$ and forms a logarithmic spiral transverse to the magnetic field. In terms of their structure, these 2D vortices are topological solitons like the 1D domain (7.84).

It is of interest to find out whether 3D soliton vortices in the magnetization can exist. They might correspond to functions of the form $\theta = \theta(r,z)$ along the anisotropic axis in an external magnetic field. Substituting these expressions into (7.80) and (7.81), we obtain an equation for θ:

$$l^2(r^{-1}\partial_r r\,\partial_r + \partial_z^2)\,\theta = [(b + l^2/r^2)\cos\theta + \omega_H/\omega_0]\sin\theta. \tag{7.87}$$

Equation (7.87) seems to have a smooth solution in 3D space that satisfies the following boundary conditions: θ goes to zero with distance from the origin, while when $r = 0$ at the axis of symmetry, we have $\theta = 0$ and $\nabla\theta = 0$, in contrast to (7.85). The toroidal and poloidal components of **m** in this solution do not depend on the angle χ. No one has yet attempted a numerical solution of (7.87). A spherically symmetrical numerical solution in which $\varphi = \text{const}$ is known and is described by (7.87) with $l = 0$ on the right-hand side (Kosevich, Ivanov and Kovalev, 1983).

8. Conservation of helicity and the Lyapunov stability of MHD structures

8.1. CONSERVATION OF MAGNETIC HELICITY

To heat a plasma, one must insulate it from the walls by means of a magnetic field created by external conductors and the current in the plasma. The confined plasma can be treated as a magnetohydrodynamic (MHD) structure. The most promising configuration is thought to be a torus in which the winding magnetic field lines form embedded magnetic surfaces. The principal difficulty to be overcome in creating such structures is to make them stable. Until recently, sufficient conditions for stable confinement had not been stated, even in the approximation of ideal MHD.

The spectral method is used in most papers to investigate plasma stability. To do this, one linearizes the MHD equations around the equilibrium state and investigates the frequencies of linear oscillations. The energy principle (Bateman, 1979) is a simpler version of this method. This approach yields sufficient conditions for a given configuration to be unstable or for the absence of perturbations that would grow exponentially in time. This method, however, cannot be used to establish that there are not perturbations that grow according to a power law (nonlinear instability).

The sufficient condition for stability is fulfilled if the nonlinear system under investigation has a Lyapunov functional that is positive definite for an equilibrium configuration (see Appendix 2). This condition also places some constraints on the equilibrium configuration to which it can be applied. For the case of an ideal fluid, one should require that the equilibrium be an extremum of one of the first integrals of motion. If the second variation of the integral in question happens to be strictly positive at this point, then the state is stable, and the integral is a Lyapunov functional.

The first integrals of the MHD equations that we know yield a rather restricted class of equilibria. For example, there is an equilibrium with pressure gradient equal to zero (Taylor, 1974). Equilibria with a finite pressure gradient have been studied by Gordin and Petviashvili (1987, 1989). They found an integral in the form of a functional series, that is, in a form that involves an arbitrary function. This integral is derived from the MHD equations for an ideal plasma

$$\rho d_t \mathbf{v} = -\nabla p + (4\pi)^{-1}(\mathrm{curl}\ \mathbf{B}\times\mathbf{B}), \tag{8.1}$$

$$\partial_t \rho + \mathrm{div}\ \rho\mathbf{v} = 0, \tag{8.2}$$

$$\partial_t p^{1/\gamma} + \mathrm{div}\ \mathbf{v}p^{1/\gamma} = 0, \tag{8.3}$$

and

$$\partial_t \mathbf{B} = \mathrm{curl}\ (\mathbf{v}\times\mathbf{B}), \quad \mathbf{B} = \mathrm{curl}\ \mathbf{A}. \tag{8.4}$$

Here it is assumed that the pressure p varies adiabatically with a ratio of specific heats γ and $\mathbf{A}$ is the vector potential of the magnetic field. The electric field in media with infinite conductivity is given by the relativistic formula $\mathbf{E} = (\mathbf{B}\times\mathbf{v})/c$. Note that it is independent of the potential part of $\mathbf{A}$. Equation (8.4) yields an equation for the vector potential,

$$\partial_t \mathbf{A} = \mathbf{v}\times\mathbf{B} + \nabla\psi, \tag{8.5}$$

where ψ is an arbitrary function. The choice of ψ determines the gauge of $\mathbf{A}$. In the Coulomb calibration, $\mathrm{div}\ \mathbf{A} = 0$ and $\nabla^2\psi = \mathrm{div}\ (\mathbf{v}\times\mathbf{B})$. Gordin and Petviashvili (1987) suggest the gauge $\psi = -\mathbf{v}\cdot\mathbf{A}$, under which (8.5) becomes

$$\partial_t \mathbf{A} = \mathbf{v}\times\mathbf{B} - \nabla(\mathbf{v}\cdot\mathbf{A}). \tag{8.6}$$

We now introduce the magnetic helicity $s = \mathbf{A}\cdot\mathbf{B}$. From this and (8.4) and (8.6), we obtain

$$\partial_t s + \mathrm{div}\ \mathbf{v}s = 0. \tag{8.7}$$

The helicity is conserved locally for the gauge chosen here. This gives a new integral in the form of the functional series

$$I_{f,D} = \int_D sf\,(\rho/s,\, p^{1/\gamma}/s)\, d^3x, \tag{8.8}$$

where D is an arbitrary fluid volume, moving at velocity **v**. We now list other, well-known MHD integrals:

$$W = \int [\,\rho v^2/2 + p/(\gamma - 1) + B^2/8\pi\,]\, d^3x, \tag{8.9}$$

$$\mathbf{P} = \int \rho \mathbf{v}\, d^3x, \tag{8.10}$$

$$\mathbf{M} = \int \rho(\mathbf{r}\times\mathbf{v}) d^3x, \tag{8.11}$$

$$C = \int \rho(\mathbf{v}\cdot\mathbf{B}) d^3x, \tag{8.12}$$

$$\mathrm{T} = \oint (\mathbf{A}\cdot\mathbf{n}) d\sigma, \tag{8.13}$$

where **n** is the unit vector normal to the plasma boundary. The integral in (8.13) is taken along the boundary surface of the plasma. We note that, in general, if any vector function **a** satisfies the equation $\partial_t \mathbf{a} + \nabla(\mathbf{v}\cdot\mathbf{a}) = \mathbf{v}\times\mathrm{curl}\,\mathbf{a}$, then the scalar product $h = \mathbf{a}\cdot\mathbf{B}$ satisfies the equation of continuity $\partial_t h + \mathrm{div}\,\mathbf{v}h = 0$ for **B** given by (8.4) with arbitrary **v**.

8.2. LYAPUNOV STABILITY OF MHD STRUCTURES

We now investigate structures at rest with $\mathbf{v} = 0$, which are of the greatest interest for applications. One cannot investigate all of them using the Lyapunov method. Bearing this in mind, we write the Lyapunov functional as follows (Gordin and Petviashvili, 1987, 1989):

$$L = W + I + \lambda T, \tag{8.14}$$

where λ is a Lagrange multiplier. Here the integral I is chosen to have the simplified form

$$I = (4\pi)^{-1} \int sf(\mu) d^3x, \quad \mu \equiv p^{1/\gamma}/s, \tag{8.15}$$

where f is an arbitrary function of a single argument. When we vary L, we assume the variations in the velocity, pressure, and vector potential are independent. Equilibria that can be investigated for stability by the Lyapunov method are determined from the Euler equation $\delta L = 0$. The variation of L with respect to **v** gives

$$\rho \mathbf{v} = 0. \tag{8.16}$$

When L is varied with respect to the pressure p, the result is

$$4\pi p^{\sigma} + \sigma \partial_{\mu} f = 0, \quad \sigma \equiv (\gamma - 1)/\gamma. \tag{8.17}$$

The variation with respect to A gives (Gordin and Petviashvili, 1987)

$$\text{curl } \mathbf{B} + 2\,\theta\,\mathbf{B} + \nabla\theta \times \mathbf{A} = 0, \quad \theta \equiv f - \mu\partial_{\mu} f. \tag{8.18}$$

It is also necessary that the surface integral obtained as a result of integrating by parts should vanish for any variation δA. This yields a condition at the plasma boundary,

$$(\mathbf{n}\cdot\mathbf{B})\Big|_{\partial D} = 0. \tag{8.19}$$

Here ∂D denotes the boundary of the volume D. Two cases are possible: (1) the region D filled with plasma is surrounded by a metal shell with infinite conductivity; (2) there is a vacuum region between D and the shell within which the integral in the functional (8.15) is not taken and where $f \equiv 0$. Case (2) can give rise, apart from (8.19), to extra boundary conditions owing to the variation of the surface ∂D of the region D. In the first case, thermal isolation requires that the following conditions hold at the plasma-shell boundary:

$$p\Big|_{\partial D} = 0, \quad \partial_n p\Big|_{\partial D} = 0. \tag{8.20}$$

This means that the pressure and the heat flux are zero at the boundary. Equations (8.17)–(8.19) define equilibrium configurations that can be investigated for stability by the Lyapunov method. It is sufficient for stability that $\delta^2 L > 0$. Applying the operation ∇ to (8.17) and div to (8.18), we get

$$4\pi\nabla p = s\nabla\theta, \quad \mathbf{B}\cdot\nabla\theta = 0. \tag{8.21}$$

Now, multiplying (8.18) by **B** vectorially with the aid of (8.21), we get

$$4\pi \nabla p = \text{curl}\, \mathbf{B} \times \mathbf{B}. \tag{8.22}$$

The scalar product of (8.18) with **A** yields

$$(\mathbf{A} \cdot \mathbf{j}) + 2\, s\, \theta = 0. \tag{8.23}$$

Thus in addition to the obvious condition (8.22) equilibrium also requires (8.23) in order to investigate it for stability. Also, as can be seen from (8.17), the pressure is a function of helicity.

An MHD structure is stable, if the second variation $\delta^2 L$ is strictly positive. The latter can be written as the sum

$$\delta^2 L = l_V + l_D + l_{\partial D}, \tag{8.24}$$

where the term l_D corresponds to the volume occupied by the plasma, $l_{\partial D}$ is a term associated with the variation of the boundary, l_V corresponds to the vacuum part between the shell and the plasma. From (8.14) we get

$$l_D = \int_D \left\{ \frac{\delta B^2}{8\,\pi} + \frac{1}{2\,\gamma} \left(\frac{\delta p}{p} \right)^2 + \frac{\varrho\, \delta v^2}{2} + \theta s_2 \right. \tag{8.25}$$

$$\left. - \frac{\theta' p^{1/\gamma}}{2} \left(\frac{1}{\gamma} \frac{\delta p}{p} - \frac{s_1}{s} \right)^2 \right\} d^3 x,$$

where $s_1 \equiv \mathbf{A} \cdot \delta \mathbf{B} + \mathbf{B} \cdot \delta \mathbf{A}$, $s_2 \equiv \delta \mathbf{A} \cdot \delta \mathbf{B}$, $\delta \mathbf{B} = \text{curl}\, \delta \mathbf{A}$,

$$l_V = \frac{1}{8\,\pi} \int_V (\delta B)^2 d^3 x, \tag{8.26}$$

and

$$l_{\partial D} = \frac{1}{2} \oint_{\partial D} \xi^2 \left\{ \partial_n (s\, f) + \left[\partial_n \left(\frac{B^2}{8\,\pi} + \frac{p}{\gamma - 1} \right) \right] \right\} d^2 x. \tag{8.27}$$

Here a prime denotes the derivative with respect to the arguments, ξ is the displacement of the boundary along the normal n, and square brackets denote the difference between the extreme values of a function inside and outside the region ∂D. When plasma is in immediate contact with the shell, we have $l_V = l_{\partial D} = 0$.

It can be shown that in order for the functional (8.25) to be positive definite in δp, one must have

$$\gamma p^{\sigma} > \partial_\mu \theta. \tag{8.28}$$

Under this condition one can obtain the following lower bound estimate for l_D:

$$l_D \geq \int_D \left\{ \frac{\delta B^2}{8\pi} + \theta s_2 + T s_1^2 \right\} d^3x, \tag{8.29}$$

where

$$T \equiv \frac{f''}{f'' + \gamma s p^{1-2/\gamma}} \frac{\gamma p}{2s^2}. \tag{8.30}$$

For all δA there is a δP that makes (8.29) an equality. For (8.29) to be positive definite one must have

$$\frac{f''}{f'' + \gamma s p^{1-2/\gamma}} > 0, \tag{8.31}$$

which, when combined with (8.28), yields a necessary condition for positive definiteness:

$$sf'' > 0. \tag{8.32}$$

Using (8.17), we pass from $\theta(\mu)$ to the function $u(s)$ using the formula

$$\theta(\mu) = (4\pi)^{-1} \partial_s u(s), \quad p = p(s). \tag{8.33}$$

We then obtain

$$4\pi p = \int s \partial_s^2 u ds, \tag{8.34}$$

and

$$T = (4\pi)^{-1}\partial_s^2 u. \tag{8.35}$$

The inequalities (8.31) and (8.32) then reduce to the simple form

$$0 < sp'(s) \leq \gamma p. \tag{8.36}$$

All these conditions can be derived from the reduced functional (Gordin and Petviashvili, 1987):

$$L_1 = \int_D [B^2/2 + u(s)]d^3x. \tag{8.37}$$

The equilibrium and stability conditions are derived from this in the form $\delta L_1 = 0$ and $\delta^2 L_1 > 0$, which is equivalent to the conditions for strict positive definiteness of $\delta^2 L$. We shall further choose the function $u(s)$ so that the reduced functional will be strictly positive definite and the inequality (8.36) is satisfied by p as defined in (8.34).

Varying the functional L_1 subject to the boundary condition in A that $\mathbf{n}\cdot\mathbf{B}\big|_{\partial D} = 0$, we obtain an Euler equation which is equivalent to (8.18)

$$\text{curl }\mathbf{B} + 2u'\cdot\mathbf{B} + \nabla u'\times\mathbf{A} = 0 \tag{8.38}$$

The boundary conditions (8.20) and (8.18) imply that

$$s\big|_{\partial D} = 0, \quad \nabla s\big|_{\partial D} = 0 \tag{8.39}$$

In order to verify the strict positive definiteness of the second variation

$$\delta^2 L_1 = \int\left[\frac{\delta B^2}{2} + u'\cdot s_2 + \frac{u''\cdot s_1^2}{2}\right]d^3x, \tag{8.40}$$

we must make the choice of the norm in the space of the perturbations $\delta\mathbf{A}$ more precise. Here we choose the norm to be explicitly dependent on the extremal, near which we are examining the variation for strict positive definiteness:

$$\delta^2 L_1 \geq c\,\|\delta A\|^2, \quad \text{where } c > 0 \tag{8.41}$$

In that case we take

$$\| \delta A \|^2 = \alpha \, \| \delta B \|^2 L_2 \,. \tag{8.42}$$

Here $\alpha = \alpha(s)$ is chosen for convenience and we factorize the $L_2(\delta \mathrm{A})$ space with respect to the kernel of this functional.

The functional (8.40) has minimum at the extremal point with respect to δA which is determined by the Jacobi equation

$$\mathrm{curl}\, \delta \mathrm{B} + 2u'\delta \mathrm{B} + \nabla u' \times \delta \mathrm{A} + u'' s_1 \mathrm{B} + \mathrm{curl}\,(u'' s_1 \mathrm{A}) = 0. \tag{8.43}$$

An investigation of whether $\delta^2 L_1$ is strictly positive definite can be carried out numerically if the shell and extremal of A have a symmetry. This been done by Gordin and Petviashvili (1989). They demonstrated the stability of "stabilized pinch" plasma configurations. In these configurations the plasma is confined by an axisymmetric toroidal shell where the poloidal magnetic field is of the same order as the toroidal field. The safety factor satisfies the condition $0 < q < 1$ and decreases monotonically from the magnetic axis to plasma boundary without changing sign. These devices include the well known ZETA machine. At present, the machine that is closest to such a configuration with an "ultra-low q" is Repute-1 at the University of Tokyo, for which the relative plasma pressure is 5–15%. Murakami *et al.* (1988) have noted that instabilities in that machine are caused by dissipative effects which lead to a decrease in the current at the plasma boundary. This violates the condition of a monotonic decrease in q, as q rises at the plasma boundary. Therefore, additional measures to maintain the plasma current at the boundary region may make the plasma stable in the Lyapunov sense in machines of this type.

Appendix 1

SUMMARY OF SIMPLIFIED (MODEL) EQUATIONS

An ideal plasma is described by the Vlasov kinetic equation with self-consistent fields. This set of equations describes many different oscillation modes. Plasma oscillations involve the excitation of collective degrees of freedom in which the particles participate consistently with one another. For this reason, they can be described satisfactorily by hydrodynamic equations derived from the Vlasov equations. Further simplifications are possible with a view to isolating a given mode or pair of modes that interact with one another. This is done by expanding in terms of the small parameters contained in these equations. The result is a simplified equation (frequently called a model equation) with a small number of variables that clearly describes the phenomenon under consideration with sufficient accuracy.

Wave equations are usually divided into acoustic and optical types. The equations describing optical oscillations involve the large parameter ω_0 which has the sense of a typical oscillation frequency. Acoustic equations include a parameter c_s which has the significance of a propagation velocity. In the short wavelength limit, both types of plasma oscillations are significantly influenced by dispersion, which is characterized by the parameter r_D. The small quantities that are used to simplify the equations are typically this parameter and the wave amplitude. Korteweg and de Vries suggested that these must be treated as small quantities of the same order in simplifying the equations. This fruitful idea has been used to obtain many different equations. There is one more parameter in an inhomogeneous medium, namely, the ratio of packet size to inhomogeneity size. Simplification according to this scheme usually reduces acoustic equations to the Korteweg–de Vries equation, in which the nonlinearity is quadratic in the amplitude

$$\partial_t \Phi + \Phi \partial_z \Phi + \partial_z^3 \Phi = 0, \tag{1*.1}$$

while optical equations are reduced to the nonlinear Schrödinger equation with a cubic nonlinearity

$$i\partial_t E + \partial_z^2 E \pm |E|^2 E = 0. \tag{1*.2}$$

These equations generalize differently for 2D or 3D packets, depending on whether the medium is isotropic or otherwise. For an isotropic medium, an acoustic equation usually reduces to the Kadomtsev–Petviashvili equation (Kadomtsev and Petviashvili, 1970)

$$\partial_z(\partial_t \Phi + \Phi \partial_z \Phi + \partial_z^3 \Phi) = \pm \nabla_\perp^2 \Phi. \tag{1*.3}$$

In a magnetized plasma, the equation for a 3D ion-acoustic packet travelling along the magnetic field has the form (Zakharov and Kuznetsov, 1974)

$$\partial_t \Phi + \Phi \partial_z \Phi + \partial_z^3 \Phi = -\nabla_\perp^2 \partial_z \Phi. \tag{1*.4}$$

The evolution of a 3D Langmuir wave packet, which is of the optical type, is described by Zakharov's equation for an isotropic plasma (Zakharov, 1972)

$$\nabla^2(i\partial_t \Phi + \nabla^2 \Phi) + \mathrm{div}(|\nabla \Phi|^2 \nabla \Phi) = 0. \tag{1*.5}$$

When a magnetic field is present, the same packet travelling along the field obeys the equation (Petviashvili, 1975)

$$\partial_z^2(i\partial_t E + \partial_z^2 E + |E|^2 E) = \pm \nabla_\perp^2 E. \tag{1*.6}$$

The propagation of a 1D packet of fast magnetosonic and Alfvén waves along a magnetic field is described by the equation (Mio *et al.*, 1976)

$$\partial_t b + i\partial_z^2 b + \partial_z |b|^2 b = 0. \tag{1*.7}$$

The simplified equation for a packet of fast magnetosonic waves propagating at a large angle to the magnetic field has the form (Manin and Petviashvili, 1983; Mikhailovskii *et al.*, 1985)

$$\partial_y(\partial_t b + b\partial_y b + \alpha\partial_y^3 b) = -\nabla_\perp^2 b + \partial_z\partial_y^3 b + \partial_y^2\partial_z^2 b. \qquad (1^*.8)$$

A simplification that only preserves the main features of a phenomenon usually results in equations that do not contain parameters. Sometimes, as in the present case, the simplified equation does contain a dimensionless parameter α which either cannot be removed by a similarity transformation or is kept for convenience.

For propagation at a large angle to the magnetic field, the velocity of low-frequency ion-acoustic waves approaches the electron drift velocity u. In this case they are well described by the 2D equation derived by Hasegawa and Mima (1978), which coincides with that obtained by Charney (1947) for atmosphere vortices whose typical size is large compared with the depth of the atmosphere and whose frequencies are small compared to the planet's rate of rotation (quasigeostrophic vortices):

$$\partial_t(\Phi - \nabla_\perp^2\Phi) + u\partial_y\Phi = \{\Phi, \nabla_\perp^2\Phi\}. \qquad (1^*.9)$$

A nonlinearity appears in the form of the Jacobian

$$\{\Phi, \psi\} \equiv \partial_x\Phi\partial_y\psi - \partial_y\Phi\partial_x\psi. \qquad (1^*.10)$$

In a homogeneous plasma u is equal to zero. For comparison we note that 2D vortices in an incompressible fluid are described by the equation

$$\partial_t\nabla_\perp^2\Phi + \{\Phi, \nabla_\perp^2\Phi\} \equiv d_t\nabla_\perp^2\Phi = 0. \qquad (1^*.11)$$

This same equation describes a Langmuir wave packet propagating at a large angle to the magnetic field, provided the ions can be treated as stationary. It has been assumed in deriving (1*.9) for an atmosphere that the pressure Φ is a function of the density N only. When this is not so, synoptic perturbations in an atmosphere are described by the equations for a shallow atmosphere (Petviashvili and Pokhotelov, 1988)

$$\begin{aligned} &\partial_t(\Phi - \nabla_\perp^2\Phi) + u_1\partial_y\Phi = \{\Phi, \nabla_\perp^2\Phi\} + \{\Phi, N\}, \\ &\partial_t N + \alpha\{\Phi, N\} + u_2\partial_y\Phi = 0, \end{aligned} \qquad (1^*.12)$$

Synoptic perturbations that essentially depend on the vertical coordinate z are described by the equation

$$\partial_t(\Phi - \nabla^2\Phi) + u\partial_y\Phi = \{\Phi, \nabla^2\Phi\}. \qquad (1^*.13)$$

If the electron temperature in a plasma is not uniform, then (1*.9) includes an additional nonlinearity of the Korteweg–de Vries type (Petviashvili, 1977):

$$\partial_t\Phi + \partial_y(u\Phi + \nabla_\perp^2\Phi + \alpha\Phi^2) = \{\Phi, \nabla_\perp^2\Phi\}. \qquad (1.^*14)$$

This same equation provides a description of large-scale atmospheric perturbations ($\nabla_\perp^2\Phi \ll \Phi$).

Alfvén waves travelling nearly perpendicular to the magnetic field are described by a set of equations derived by Kadomtsev and Pogutse (1973):

$$\begin{aligned} &d_t\nabla_\perp^2\Phi + d_z J = 0, \quad d_z \equiv \partial_z - \{A, \dots\}, \\ &\partial_t A + d_z\Phi = 0, \qquad J \equiv \nabla_\perp^2 A, \end{aligned} \qquad (1^*.15)$$

The propagation velocity may sometimes be so small as to require inclusion of the diamagnetic electron drift velocity in (1*.15). One should also include the dispersion corrections associated with the electric field parallel to the magnetic field. These equations are then replaced by the set (Petviashvili and Pokhotelov, 1985)

$$\begin{aligned} &d_t\nabla_\perp^2\Phi + d_z J = -\mathrm{div}\,\{P, \nabla_\perp\Phi\}, \\ &\partial_t A = d_z(N - \Phi), \\ &d_t N + d_z J = 0, \quad d_t P = 0. \end{aligned} \qquad (1^*.16)$$

Up to now we have assumed the packet size to be large compared to the ion Larmor radius r_{Bi}.

When the typical size of an Alfvén wave packet is small compared to r_{Bi}, (1*.16) is replaced by

$$\left.\begin{aligned} &\tau\partial_t\Phi + u\partial_y\Phi = d_z\nabla_\perp^2 A, \\ &\partial_t A - u\partial_y A = -(1+\tau)d_z\Phi. \end{aligned}\right\} \qquad (1^*.17)$$

(see Petviashvili and Pogutse, 1984). It is assumed in (1*.16) and (1*.17) that the plasma pressure is sufficiently high that the Alfvén speed is below the thermal electron speed. Otherwise, an Alfvén wave packet is described by the equations (Kaladze, Marchenko, Pokhotelov and Pet-viashvili, 1987)

$$d_t \nabla_\perp^2 \Phi + d_z J = 0,$$
$$\partial_t A - \alpha d_t J = d_z(N - \Phi), \qquad (1^*.18)$$
$$d_t N + d_z J = 0.$$

In Alfvén waves the magnetic field lines experience transverse oscillations described by A, the longitudinal component of the vector potential. Simultaneous oscillations exist in the electric potential Φ and lead to vortical plasma motion nearly perpendicular to the magnetic field. Oscillations strictly perpendicular to the magnetic field are described by the equations for flute waves (Petviashvili and Pogutse, 1986):

$$d_t \nabla_\perp^2 \Phi + \operatorname{div} \{N, \nabla_\perp \Phi\} = \alpha \partial_x N, \quad d_t N = 0. \qquad (1^*.19)$$

Appendix 2

LYAPUNOV STABILITY OF EQUILIBRIUM SOLUTIONS FOR EVOLVING SYSTEMS

In 1914 Lyapunov developed the theory of stability for solutions of sets of ordinary differential equations. The general theory is fairly complex. Below we quote some results from this theory relating to the stability of the equilibrium state for autonomous systems, that is, systems of the type

$$d_t x = f(x), \tag{2*.1}$$

where $x = (x_1, \ldots, x_n)$ is an n-dimensional vector and $f = (f_1, \ldots, f_n)$ is an n-dimensional vector function of x that does not depend explicitly on the time.

The system of equations (2*.1) describes the path of the point x in n-dimensional phase space. Of special interest are the points $x = x_0$ of the phase space where $f(x_0) = 0$, which are known as equilibrium points. If a phase point is in a neighbourhood of x_0, three cases can arise: (a) all points of the phase space approach x_0 as time goes on; (*b*) all points of the space are oscillating around x_0 and stay within a short distance of it; and (*c*) some or all paths move away from the equilibrium point. In the first two cases the equilibrium is called stable in the sense of Lyapunov. He showed that then there is a scalar function $L(x)$ that satisfies the following conditions in neighbourhood of x_0:

(1) $L(x)$ has continuous derivatives,

(2) $L(x_0) = 0$,

(3) $L(x) > 0$ for $x \neq x_0$,

(4) During the evolution of x, $L(x)$ does not increase because of (2*.1). The last condition means that

$$d_t L(x) = \sum_{k=1}^{n} \partial_{x_k} L(x) d_t x_k = \sum_{k=1}^{n} \partial_{x_k} L(x) f_k(x) \leq 0. \qquad (2^*.2)$$

The function $L(x)$ is called the Lyapunov function.

Thus, an equilibrium point of the autonomous system (2*.1) is stable, provided a Lyapunov function exists near it. This theorem is now also used for partial differential equations, with the n-dimensional vector replaced by an infinite dimensional vector. A function corresponds to an infinite dimensional vector in partial differential equations. The vector components are numbered by the spatial coordinates. The Lyapunov function becomes a functional and the derivatives in (2*.2) become variational derivatives. The condition (2*.2) is, in particular, satisfied, if L is a first integral of the system. Because of the smoothness of L, this implies that

$$\partial_x L\big|_{x_0} = 0 \text{ or } \delta L\big|_{x_0} = 0. \qquad (2^*.3)$$

This relation combined with $f(x_0) = 0$ is used in practice to choose possible stability points from among all the equilibrium points x_0. The stability of a stationary point is finally established by appealing to condition 3, which assumes the form

$$\delta^2 L\big|_{x_0} > 0 \qquad (2^*.4)$$

for partial differential equations.

Under these conditions, the point $x(t)$ will never leave the vicinity of an equilibrium point, so that x_0 is a stationary point with Lyapunov stability. The spectral method is sometimes used instead of the Lyapunov method, as it is the simpler of the two methods. It involves investigating the spectrum of linear oscillation frequencies relative to the equilibrium point. According to this method, an equilibrium is stable if all the frequencies decay and is unstable if at least one grows. However, when at least one frequency is neutral (real), while the remaining frequencies decay, this method does not work. We illustrate this for the linear system

$$\dot{x} = -y + (r - \nu)x, \quad \dot{y} = x + (r - \nu)y, \quad r^2 = x^2 + y^2. \qquad (2^*.5)$$

The frequencies relative to the point of rest (equilibrium) $r = 0$ are $\omega = \pm 1 - i\upsilon$. If $\nu > 0$, then the equilibrium is stable, and if $\nu < 0$, it is unstable, while if $\nu = 0$ (real frequencies) further investigation is needed. This is easy to do in this particular case. The exact solution for $\nu = 0$ is

$$r = 1/(t_0 - t). \tag{2*.6}$$

From (2*.6) one can see that a power-law growth (nonlinear instability) is not detected by the spectral method. Frequencies around an equilibrium point are usually real in ordinary hydrodynamics or MHD with no dissipation. Hence, the spectral method is insufficient for determining stability and the Lyapunov method is more useful.

REFERENCES

Abramyan L.A., Stepanyants Yu.A. (1985), The structure of two-dimensional solitons in media with anomalously small dispersion, *Sov. Phys. JETP*, **61**, 963.

Abrashkin A.A., Yakubovich E.I. (1984), Planar rotational flows of an ideal fluid, *Sov. Phys. Dokl.*, 29, (No. 5), 370-371.

Akasofu S.I., Chapman S. (1972), *Solar-Terrestrial Physics*, Oxford, Clarendon Press.

Alexander J.K., Kaiser M.L. (1977). Terrestrial kilometric radiation. 2. Emission from the magnetospheric cusp and dayside magnetosheath, *J.Geophys. Res.*, **82**, (No. 1), 98-104.

Antipov S.V., Nezlin M.V., Snezhkin E.N., and Trubnikov A.S. (1981), Rossby solitons, *Sov. Phys. JETP Letters*, 33, (No. 7), 351-355.

Antonova R.A., Zhvania B.P., Lominadze D.G., Nanobashvili D.I. and Petviashvili V.I. (1983), Drift solitons in a shallow rotating fluid, *Sov. Phys. JETP Lett.*, 37, (No. 11), 651-655.

Antonova R.A., Zhvania B.P, Lominadze D.G., Nanobachvili D.I. and Petviashvili V.I. (1987), Use of rapidly rotating shallow water to model vortices in a magnetized plasma, Sov. J. Plasma Phys., 13, (No. 11) 765-767.

Arnold V.I. (1978), Mathematical methods of Classical mechanics, Graduate texts in mathematics, 60, Berlin, Springer.

Artsimovich L.A., Sagdeev R.Z. (1979), *Plasma Physics for Physicists*, N.Y., Harwood Academic Press.

Askar'yan G.A. (1962), Effects of the gradient of a strong electromagnetic beam on electrons and atoms, *Sov. Phys. JETP Letters*, 15, (No. 6), 1088-1090.

Baines P.A. (1983), A survey of blocking mechanisms with application to the Australian region, *Austral. Meteorol. Mag.*, 31, (No. 1), 27-36.

Bateman G. (1979), *MHD Instabilities*, Cambridge, Massachusetts and London, MIT Press.

Belcher J.W., Davis L.Jr. (1971), Large-amplitude Alfvén waves in the Interplanetary Medium, 2, *J. Geophys. Res.*, 76, 3534.

Bel'kov S.A., Tsytovich V.N. (1979), Modulation excitation of magnetic fields, *Sov. Phys. JETP*, 49, (No. 4), 656-661.

Berestov A.L. (1981). Some new solutions for Rossby solitons, *Isvestia Akademii Nauk, Fizika atmosphery i okeana*, 17, (No. 1); 82-87.

Bergmann R. (1984), Electrostatic ion (hydrogen) cyclotron and ion acoustic wave instabilities in regions of upward field-aligned current and upward ion beams, *J.Geophys. Res.*, 89, (No. A2), 953-968.

Bordag L.A., Its A.R., Matveev V.B. et all (1977), *Phys. Lett.*, 63A, (No. 3), 205.

Borodachev L. N., Nekrasov A. K. (1984), Diamagnetic instability of cyclotron waves in plasma, *Vestnik Moscovskogo Universiteta. Seriya 3. Fizica. Astronomiya*, 25, (No. 5), 91-98 (in Russian).

Braginsky S.I. (1965), Transport processes in plasmas, In: *Reviews of Plasma Physics*, 4, Ed. by M.A.Leontovich N.Y. Consultants Bureau, 205-311.

Budker G.I. (1965), *CERN Symposium*, 1, 68-71.

Charney J.G. (1947), On the scale of atmospheric motions, *Geophys. Publ.*, 17, (No. 2), 17-20.

Chen F. (1984), *Introduction to Plasma Physics and Controlled Fusion*, N.Y., London, Plenum Press.

Cheung P.Y., Wong A.Y. (1985), Periodic collapse and long-time evolution of strong Langmuir turbulence, *Phys. Rev. Lett.*, 55, (No. 18), 1880-1883.

Coleman P.J. (1968), *J.Geophys. Res.*, 153, 371-378.

Chmyrev V.M., Bilichenko S.V., Pokhotelov O.A. and Marchenko V.A. (1988), Alfvén vorticies and related phenomena in the ionsphere and magnetosphere, *Physica Scripta*, 38, 841-854.

Cole K.D., Pokhotelov O.A. (1980), Cyclotron solitons source of Earth's kilometric radiation, *Plasma Physics*, 22, 595-608.

Derrick G.H. (1964), Comments on nonlinear wave equations as models for elementary particles, *J.Math.Phys.*, 5, (No. 9), 1252-1254.

Dobrovolny M., Mengeney A., Veltry P. (1980), Properties of magnetohydrodynamic turbulence in Solar wind, *Astron. and Astrophysics*, 83, 26-32.

Dowling T.E., Ingersoll A.I. (1988), Potential vorticity and layer thickness variations in the flow around Jupiter's GRS and white oval BC, *J. Atmospheric Sci.*, 45, (No. 8), 1380-1396.

Dryuma V.S. (1974), Analytic solution of the two-dimensional Korteweg-de-Vries (KdV) equation, *Sov. Phys. JETP Lett.*, 19, (No. 12), 387-388.

Drummond W.E., Rosenbluth M.N. (1962), Anomalous diffusion arising from microinstabilities in a plasma, *Phys. Fluids*, 5, (No. 12), 1507-1513.

Feldstein A.Ya., Petviashvili N.V. (1989). Amplification of synoptic vortices localized in the thermosphere, *Izvestiya Vuzov Radiofizika*, 32, (No. 11), 1315-1319 (in Russian).

Filippov N.V. (1983), Plasma-focus experiments at the Kurchatov Institute, Moscow (review), *Sov. J. Plasma Physics*, 9, (No. 1), 14-25.

Flierl G.R. (1979), Baroclinic solitary waves, *Dynamics Atmosph. Oceans*, 3, (No. 1), 15-38.

Flierl G.R., Larichev V.D., McWilliams J.C., Reznik G.M. (1980), The dynamics of baroclinic and barotropic solitary eddies, *Dynamics Atmosph. Oceans*, 5, (No. 1), 1-41.

Galeev A.A. (1963), Instability theory for a low-pressure inhomogeneous plasma in a strong magnetic field, *Sov. Phys. JETP*, 17, (No. 6), 1292-1301.

Galeev A.A., Sagdeev R.Z. (1979), Nonlinear plasma theory, In: *Reviews of Plasma Physics*, 7, Ed. by M.A.Leontovich, Consultants Bureau, NY, London, 1-180.

Gardner C.S., Green J.M., Kruskal M.D., Miura R.M. (1967), Method for solving the Korteveg de-Vries Equation, *Phys. Rev. Lett.*, 19, 1095-1097.

Gordin V.A., Petviashvili V.I. (1987), Gauge of the vector potential Lyapunov - stable MHD equilibria, *Sov. J. Plasma Phys.*, 13, (No. 7), 509-511.

Gordin V.A., Petviashvili V.I. (1989), Lyapunov stability of MHD equilibrium of a plasma with nonvanishing pressure, *Sov. Phys. JETP*, 68, (No. 5), 988-994.

Gorshkov K.A., Mirnov V.A., Sergeev A.M. (1983), Bounded stationary solitary structures, In: *Nonlinear waves*, Moscow, Nauka, 112-128 (in Russian)

Gurnett D.A. (1974), The Earth as a radio source: terrestrial kilometric radiation, *J.Geophys. Res.*, 79, (No. 28), 4227-4238.

Hasegawa A. (1975), *Plasma Instabilities and Nonlinear Effects*, Springer-Verlag Berlin, Heidelberg, N.Y.

Hasegawa A. (1976), Particle acceleration by MHD surface wave in formation of aurora, *J.Geophys. Res.*, 18, (No. 28), 5083-5090.

Hasegawa A., Mima K. (1976), Exact solitary Alfvén wave, *Phys. Rev. Letters.*, 37, (No. 11), 690-693.

Hasegawa A., Mima K. (1978), Pseudo-three dimensional turbulence in magnetized nonuniform plasma, *Phys. Fluids*, **21**, (No. 1), 87-92.

Hasegawa A. (1985), *Advances in Physics*, **34**, 1.

Hirota R. (1973), Exact N-soliton solutions of the wave equation of long waves in shallow water and in nonlinear lattices, *J.Math.Phys.*, **19**, 810-815.

Isichenko M.B., Marnachev A.M. (1987), Nonlinear wave solutions of electron MHD in a uniform plasma, *Sov.Phys. JETP*, **66**, 7, 702-708.

Ivanov V.N., Pokhotelov O.A. (1987). Flute instability in the plasma sheath of the Earth's magnetosphere. *Sov. J. Plasma Physics*, **13**, 833-842.

Ivanov V.N., Pokhotelov O.A. (1988), Vortex tubes in a dipole magnetic field, *Sov. J. Plasma Physics*, **14**, (No. 10), 694-697.

Kadomtsev B.B. (1965), *Plasma Turbulence*, London, Academic Press.

Kadomtsev B.B. (1966), Hydrodynamic plasma stability, In: *Reviews of Plasma Physics*, 2, Ed. by M.A.Leontovich, N.Y.Consultants Bureau.

Kadomtsev B.B., Mikhailovskii A.B., Timofeev A.V. (1965), Negative energy waves in dispersive media, *Sov. Phys. JETP*, **20**, (No. 6), 1516-1518.

Kadomtsev B.B., Pogutse O.P. (1968), Electric conductivity of a plasma in a strong magnetic field, *Sov. Phys. JETP*, **26**, (No. 6), 1146-1150.

Kadomtsev B.B., Petviashvili V.I. (1970), On the stability of solitary waves in weakly dispersing media, *Sov. Phys. Dokl.*, **15**, (No. 6), 539-541.

Kadomtsev B.B., Petviashvili V.I. (1973). On acoustic turbulence, *Dokl. Academii Nauk*, **218**, (No. 4), 794-796 (in Russian).

Kadomtsev B.B., Pogutse O.P. (1974), Nonlinear helical perturbations of tokamak plasmas, *Sov. Phys. JETP*, **38**, (No. 2), 283-290.

Kadomtsev B.B. (1976), *Collective Phenomena in Plasmas*, Moscow, Nauka (in Russian).

Kadomtsev B.B., Pogutse O.P. (1984). Theory of electron transport in a strong magnetic field, *Sov. Phys. JETP Letters.*, **39**, (No. 5), 269-272.

Kadomtsev B.B. (1987), Magnetic field line reconnection, *Rep. Prog. Phys.*, **50**, 115-138.

Kaladze T.D., Petviashvili V.I., Pokhotelov O.A. (1986), Condensation of Alfvén waves into vortices in an inhomogeneous plasma, *Sov. Phys. JETP*, **64**, (No. 1), 62-66.

Kaladze T.D., Marchenko V.A., Pokhotelov O.A., Petviashvili V.I. (1987), Negative energy vortices in an inhomogeneous plasma, *Plasma Physics and Controlled Fusion*, **29**, (No. 5), 580-599.

Karpman V.I. (1975), *Nonlinear Waves in Dispersive Media*, Pergamon press, Oxford, NY, Toronto, Sydney.

Kaup D.J., Newell A.C. (1978), An exact solution for a derivative nonlinear Schrödinger equation, *J.Math. Phys.*, **19**, 798-801.

Korchagin V.I., Petviashvilii V.I. (1985), Rossby solitons in the disk of the galaxy, *Sov. Astron. Lett.*, **11**, (No2), 121-122.

Kosevich A.M., Ivanov B.A., Kovalev A.S. (1983), *Nonlinear Magnetization Waves*, Kiev, Naukova Dumka (in Russian).

Kuznetsov E.A., Musher S.L., Shafarenko A.V. (1983), Collapse of acoustic waves in media with positive dispersion, *Sov. Phys. JETP Letters.*, **37**, (No. 5), 241-244.

Kusmartsev F.V. (1984), On the classification of solitons, *Physica Scripta*, **29**, 7-11.

Kraichman R.H. (1965), Intertial range spectrum of hydromagnetic turbulence, *Phys. Fluids*, **8**, (No7), 1385-1387.

Kuznetsov E.A., Turitsyn S.K. (1982), Two-and-three dimensional solitons in weakly dispersive media, *Sov. Phys. JETP*, **55**, (No. 5), 844-847.

Laedke E.W., Spatshek K.H. (1986), Two-dimensional drift vortices and their stability, *Phys. Fluids*, **29**, (No. 1), 133-142.

Landau L.D., Lifshitz E.M. (1963), Electrodynamics of Continious Media, *Course of Theoretical Physics*, **8**, Pergamon Press.

Larichev V.D., Reznik G.M. (1976), On 2D solitary Rossby waves, *Doklady Akademii Nauk SSSR*, 231, 1077-1079. (in Russian).

Larichev V.D., Reznik M.G. (1982), Numerical experiment on the study of solitary Rossby wave collisions, *Doklady Akademii Nauk SSSR*, **264**, (No. 1), 229-233.

Lax P.D. (1968), Integrals of nonlinear equations of evolutions and soli-tary waves, *Comm. Pure Appl. Math.*, **21**, 467-472.

Longmire C.L., Rosenbluth M.N. (1957), Stability of plasmas confined by magnetic fields, *Ann. Phys.*, **1**, 120.

Lean G. (Ed.) (1985), *Radiation Doses, Effects, Risks*, United Nations Environment Program.

Leontovich M.A. (1944), On one method of the solution of problem of propagation of electromagnetic waves along the Earth's surface, *Izvestiya Akademii Nauk SSSR, Seriya Fizika*, **8**, (No. 1), 16-22.

Liewer P.C. (1985). Review of measurements of microturbulence in tokamaks and comparisons with theories of turbulence and anomalous transport, *Nuclear Fusion*, **25**, 543.

Lin C.C. (1955), *The Theory of Hydrodynamic Stability*, Cambridge University Press.

Litvak A.G. (1986), Dynamic nonlinear electromagnetic phenomena in plasmas, In: *Review of Plasma Physics*, **10**, NY - London, Consultants Bureau, ed. by M.A. Leontovich.

Manin D.Yu., Petviashvili V.I. (1984), Self-focusing of a magnetosonic wave across the magnetic field, *Sov. Phys. JETP Letters*, 38, (No. 9), 517-520.

Marnachev A.M. (1987), Localized toroidal vorticies in a Z-pinch, *Sov. J. Plasma Phys.*, 13, (No. 5), 312-316.

Marnachev A.M. (1988), On drift-Alfvén vorticies, *Fizika Plazmy*, **14**, (No. 6), 832-844. (in Russian).

Maxworthy T., Redekopp L.G. (1976). A solitary wave theory of the Great Red Spot and other observed features in the Jovian atmosphere, *Icarus*, 29, 261-271.

McMahon A.B. (1968), Discussion of paper by C.F.Kennel and R.Z.Sagdeev "Collisionless shock waves in high-β plasma, 2", *J. Geophys. Res.*, 73, 7539.

McWilliams T.C., Zabusky N.J. (1982), Interaction of isolated vortices, Geophys. Astrophys. *Fluid dynamics*, **19**, 207-227.

Mikhailovskaya L.V., Mikhailovskii A.B. (1963), Drift instability of a plasma in a helical magnetic field, *Nuclear Fusion*, **3**, 113.

Mikhailovskii A.B., Rudakov L.I. (1963), The stability of a spatially inhomogeneous plasma in a magnetic field, *Sov. Phys. JETP*, **17**, (No. 3), 621-625.

Mikhailovskii A.B. (1974), *Theory of Plasma Instabilities*, vols. 1 and 2, Consultants Bureau, NY.

Mikhailovskii A.B., Aburdzhaniya G.D., Onischenko O.G., Smolyakov A.I. (1985), The structure of the nonlinear equations of a magnetized plasma and the problem of the stability of magnetosonic solitons, *Sov. Phys. JETP*, **62**, (No. 2), 273-281.

Mio K., Ogino T., Minamy K., Takeda S. (1976), Modified nonlinear Schrödinger equation for Alfvén waves propagating along the magnetic field in cold plasmas, *J. Phys. Soc. Japan*, **41**, 265-273.

Mirnov S.V. (1985), Physical *Processes in Tokamaks*, Moscow, Energoatomizdat. (in Russian).

Moiseev S.S., Sagdeev R.Z. (1964), On the Bohm diffusion coefficient, *Sov. Phys. JETP*, **17**, (No. 2), 515-517.

Morozov A.I., Solov'ev L.S. (1980), Steady-state plasma flow in magnetic field. In: *Rev. Plasma Physics*, **8**, Consult. Bureau N.Y., 8, 1-103.

Mozer F.S., Carlson C.W., Hudson M.K. (1977), Observations of paired electrostatic shocks n the polar magnetosphere, *Phys. Rev. Lett.*, 38, (No. 6), 292-295.

Mukhovatov V.S. (1980), Tokamaks, In: "Itogi nauki i tekhniki", Seriya Fiziki Plasmy, Moscow, *VINITI*, **1**, (part 1), 6-118 (in Russian).

Murakami Y., Yoshida Z., Inoue N. (1988), Ideal MHD stability analysis for equilibria with a pitch minimum, *Nuclear Fusion*, **28**, (No. 3), 449-455.

Nekrasov A.K., Petviashvili V.I. (1979). Diamagnetism of cyclotron waves in plasmas, *Sov. Phys. JETP*, **50**, (No. 2), 305-310.

Nekrasov A.K., Petviashvili V.I. (1981), Self-focusing and three-dimensional localization of cyclotron wave traveling along a magnetic field, *Sov. J. Plasma Phys.*, **7**, (No. 5), 630-633.

Nekrasov A.K., Feigin F.Z. (1985), Magnetic nonlinearity of short-wave packets of cyclotron waves, *Sov. J. Plasma Physics*, **11**, (No. 8), 565-568.

Nekrasov A.K., Pavlenko V.P., Petviashvili V.I. (1985), Convection during rf plasma heating, *Sov. J. Plasma Physics*, **11**, (No. 10), 725-726.

Nekrasov A.K. (1986), Diamagnetic self-focusing of electromagnetic cyclotron waves propagating across a magnetic field, *Sov. J. Plasma Phys.*, **12**, (No. 8), 557-562.

Nezlin M.V. (1986), Rossby solitons (Experimental investigations and laboratory model of natural vortices of the Jovian Great Red Spot type) *Sov. Phys. Uspekhi*, **29**, (No. 9), 807-842.

Oraevsky V.N., Tasso H., Wobig H. (1969), Nonlinear drift waves in a plasma with a temperature gradient, *Proc. of ICPP and CNFR held in Novosibirsk*, CN-24/E-6, Vienna.

Ozhogin V.I., Preobrazhenskii V.Z. (1977), Effective anharmonicity of the elastic subsystem of antiferromagnets, *Sov. Phys. JETP*, **46** (No. 3), 523-529.

Ozhogin V.I., Manin D.Yu., Petviashvili V.I., Lebedev A.Yu. (1983), Self-focusing of sound wave in magnetics with high effective anharmonicity, *IEEE Transactions on Magnetics*, *Mag*-**19**, (No. 5), 1977-1979.

Pavlenko V.P., Petviashvili V.I. (1982), Band theory for the stability of nonlinear periodic waves in plasmas, *Sov. J. Plasma Phys.*, **8**, (No. 1), 117-120.

Pavlenko V.P., Petviashvili V.I. (1983), Solitary vortex in a flute instability, *Sov. J. Plasma Phys.*, **9**, (No. 5), 603-604.

Pedlosky J. (1979), *Geophysical Fluid Dynamics*, Springer-Verlag, N.Y., Heidelberg, Berlin.

Petviashvili V.I. (1964), Accoustic ion oscillations excited by an electron current, *Sov. Phys. Doklady*, **8**, (No. 12), 1218-1220.

Petviashvili V.I. (1975), Three-dimensional solitons of extraordinary and plasma waves, *Sov. J. Plasma Phys.*, **1**, (No. 1), 15-16.

Petviashvili V.I. (1976a), Equation of an extraordinary soliton, *Sov. J. Plasma Physics*, **2**, (No. 3), 257-258.

Petviashvili V.I. (1976b), HF-diamagnetism and three-dimension cyclotron solitons in a plasma, *Sov. Phys. JETP Letters.*, **23**, (No. 12), 627- 629.

Petviashvili V.I. (1977), Self-focusing of an electrostatic drift wave, Sov. *J. Plasma Phys.*, **3**, (No. 2), 150-151.

Petviashvili V.I., Pokhotelov O.A. (1977), Alfvén and magnetosonic vortices in a plasma, *Sov. Phys. JETP*, **46**, (No. 2), 260-264.

Petviashvili V.I., Tsvelodub O.Yu. (1978), Horseshoe-shaped solitons on a down-ward flowing film of viscous fluid. *Sov. Phys. Dokl.*, **23**, (No. 2), 117-118.

Petviashvili V.I., Tsvelodub O.Yu. (1980), Three-dimensional dissipative Langmuir soliton, *Sov. J. Plasma Phys.*, **6**, (No. 2), 257-258.

Petviashvili V.I., Red Spot of Jupiter and the drift soliton in a plasma (1980), *Sov. JETP Letters*, **32**, (No. 11), 619-622.

Petviashvili V.I., Pokhotelov O.A., Chudin N.V. (1982), Solitary toroidal vortices, *Sov. Phys. JETP*, **55**, (No. 6), 1056-1059.

Petviashvili V.I. (1983), Solitary vortices subject to zonal flow in a rotating atmosphere, *Sov. Astron. Letters.*, **9**, (No. 2), 137-138.

Petviashvili V.I., Pokhotelov O.A. (1983), *Vortices in shallow rotating atmosphere. In Nonlinear Waves*, Moscow, Nauka, 107-112 (in Russian).

Petviashvili V.I., Pogutse I.O. (1984), Flute solitons in a plasma with shear, *Sov. Phys. JETP Letters*, **39**, (No. 8), 437-439.

Petviashvili V.I., Pokhotelov O.A. (1985), Dipole Alfvén vortices, *Sov. Phys. JETP Letters.*, **42**, (No. 2), 54-56.

Petviashvili V.I., Pokhotelov O.A. (1986), Solitary vortices in plasmas, *Sov. J. Plasma Phys.*, **12**, (No. 9), 651-661.

Petviashvili V.I., Pogutse I.O. (1986), Drift-dissipative excitation of electron vortices in a plasma, *Sov. Phys., JETP,. Letters.*, **43**, (No. 6), 343-345.

Petviashvili V.I., Pokhotelov O.A., Stenflo L. (1986), Toroidal Alfvén solitons in a space plasma, *Sov. J. Plasma Phys.*, **12**, (No. 8), 545-547.

Petviashvili V.I., Pokhotelov O.A. (1988), The equations of shallow atmosphere, *Sov. Phys. Doklady*, **300**, (No. 4), 856-858.

Petviashvili N.V. (1988), Three-dimensional Rossby solitons in boroclinic media, *Izvestiya Akademii Nauk, Fizika atmosfery i okeana*, **24**, (No7), 776-778.

Pitaevsky L. A. (1961), Electric forces in transparent dispersive medium, *Sov. Phys. JETP*, **12**, (No. 5), 1009-1013.

Rabinovich M.I., Trubetskov D.I. (1984), *Introduction to the theory of oscillations and waves*, Moscow, Nauka (in Russian).

Rudakov L.I. (1973), Deceleration of electron beams in a plasma with high level of Langmuir turbulence, *Sov. Phys. Dokl.*, **17**, (No. 12), 1166-1167.

Russel C.T., Holtzer R.E., Smith E.J. (1970), OGO-3 observations of ELF noise in the magnetosphere, 2, The nature of the equatorial noise, *J.Geophys. Res.*, **75**, (No. 4), 755-768.

Sagdeev R.Z. (1966), Cooperative phenomena and shock waves in collectionless plasmas, In: *Reviews of Plasma Physics*, Ed. by M.A. Leontovich, 4, 23-91, Consultants Bureau, N.Y.

Sagdeev R.S., Sotnikov V.I., Shapiro V.D. and Shevchenko V.I. (1977), Contribution to the theory of magnetosonic turbulence, *Sov. Phys. JETP Letters.*, **26**, (No. 11), 582-586.

Scott A. (1970), *Active and nonlinear wave propagation in electronics*, N.Y., London, Sydney, Toronto, Wiley Interscience.

Shafranov V.D. (1966), Plasma equilibrium in a magnetic field, In: *Reviews of Plasma Physics*, Ed. by M.A.Leontovich, Consultants Bureau, N.Y., London, **2**, 103-151.

Smith B.A., Soderblom L.A., Beede R., *et al.* (1979a), The Jupiter system through the eyes of Voyager-1, *Science*, **204**, (No. 4396), 951-971.

Smith B.A., Soderblom S.L., Beebe R. *et al.* (1979b), The Galilean satellites and Jupiter: Voyager-2 imagine in science results, *Science*, **206**, (No. 4421), 927-950.

Stefanovsky A.M. (1965), Plasma electrons acceleration, *Nuclear Fusion*, **5**, (No. 3), 215-227.

Streltsov A.V., Chmyrev U.M., Pokhotelov O.A., Marchenko V.A. and Stenflo L. (1990), The formation and nonlinear evolution of convective cells in auroral plasma, *Physica Scripta*, **40**, 535-545.

Tasso H. (1967), Shock like drift waves, *Phys. Lett.*, **24A**, (No. 11), 618.

Taylor J.B. (1974), Relaxation of toroidal plasma and generation of reverse magnetic fields, *Phys. Rev. Lett.*, **33**, (No. 19), 1139-1141.

TFR Group (1975), *High-current discharges in the TFR device*, Plasma Physics and Controlled Nuclear Fusion Research, 1974, Vienna: IAEA.

Timofeev A.V. (1964), Dissipative instability of a weakly ionized inhomogeneous plasma in a uniform external magnetic field, *Sov. Phys.–Technical Phys.*, **8**, (No. 8), 682-685.

Timofeev A.V. (1971), Oscillations of inhomogeneous flows of plasma and liquids, *Sov. Phys. Uspekhi, 13*, (No. 5), 632-646.

Tserkovnikov Yu.A. (1957), Stability of plasma in a strong magnetic field, *Sov. Phys. JETP*, **5**, (No. 1), 58-64.

Turitsyn S.K., Fal'kovich G.E. (1985), Stability of magnetoelastic solitons and self-focusing of sound in antiferromagnets, *Sov. Phys. JETP*, 62 (1), 146-152.

Vakhitov N.G., Kolokolov A.A. (1973), Stationary solitons of wave equation in medium with saturating nonlinearity, *Izvestiya Vuzov Radiofizika*, 16, 1020-1028. (in Russian).

Vedenov A.A., Rudakov L.I. (1965), Interaction of waves in continuous media, *Sov. Phys.–Dokl.*, 9, (No. 12), 1073-1075.

Washimi H., Karpman V.I. (1976). The ponderomotive forse of a high-frequency electromagnetic field in a dispersive medium, *Sov. Phys. JETP*, **44**, (No. 3), 528-531.

Whitham J.B. (1974), *Linear and Nonlinear Waves*, N.Y., Wiley.

Zabusky N.J., Kruskal M.D. (1965), Interaction of solitons in a collisionless plasma and recurrence of initial states, *Phys. Rev. Letters*, **15**, (No. 6), 240-242.

Zaitsev A.A. (1983), Formation of stationary nonlinear waves by superposition of solitons, *Sov. Phys. Dokl.* **28** (9), 720-722.

Zakharov V.E. (1972), Collapse of Langmuir waves, *Sov. Phys. JETP*, **35**, (No. 5), 908-914.

Zakharov V.E., Kuznetsov E.A. (1974), Three-dimensional solitons, *Sov. Phys. JETP*, **39**, (No. 2), 285-286.

Zakharov V.E., Manakov S.V., Novikov S.P., Pitaevskii L.P. (1980), *The Theory of Solitons*, Moscow, Nauka (in Russian).

Zakharov V.E. (1984), Collapse and self-focusing of Langmuir waves. In: *Handbook of Plasma Physics*, 2, (No. 4), 81-121. North-Holland Physics Publishing, Amsterdam, Oxford, Tokyo.

Zaleskii Yu.G., Zinchenko V.I., Nazarov N.I. and Demchenko V.V. (1982), Experimental observation of nonlinear effects upon the excitation of large amplitude cyclotron waves in a plasma, *Sov. Phys. JETP Letters.*, **35**, (No. 7), 347-350.

Zasov A.V., Kyazumov G.A. (1981), Gas movement and mass distribution in the spiral galactics, *Sov. Astron. Letters.*, 7, (No. 3), 131-133.

Zastavenko L.G. (1965), *Particle like solitons of nonlinear wave equation, Prikladnaya matematika i mekhanika*, **29**, 430-439. (in Russian).

Index